Oxford Applied Mathematics
and Computing Science Series

General Editors
J. N. Buxton, R. F. Churchhouse, and A. B. Tayler

OXFORD APPLIED MATHEMATICS AND COMPUTING SCIENCE SERIES

I. Anderson: *A First Course in Combinatorial Mathematics*

D. W. Jordan and P. Smith: *Nonlinear Ordinary Differential Equations*

D. S. Jones: *Elementary Information Theory*

B. Carré: *Graphs and Networks*

A. J. Davies: *The Finite Element Method*

W. E. Williams: *Partial Differential Equations*

R. G. Garside: *The Architecture of Digital Computers*

J. C. Newby: *Mathematics for the Biological Sciences*

G. D. Smith: *Numerical Solution of Partial Differential Equations* (*Third Edition*)

J. R. Ullmann: *A Pascal Database Book*

S. Barnett and R. G. Cameron: *Introduction to Mathematical Control Theory* (*Second Edition*)

A. B. Tayler: *Mathematical Models in Applied Mechanics*

A. B. TAYLER
University of Oxford

Mathematical Models in Applied Mechanics

CLARENDON PRESS · OXFORD
1986

Oxford University Press, Walton Street, Oxford OX2 6DP
Oxford New York Toronto
Delhi Bombay Calcutta Madras Karachi
Kuala Lumpur Singapore Hong Kong Tokyo
Nairobi Dar es Salaam Cape Town
Melbourne Auckland
and associated companies in
Beirut Berlin Ibadan Nicosia

Oxford is a trade mark of Oxford University Press

Published in the United States
by Oxford University Press, New York

© *A. B. Tayler*, 1986

British Library Cataloguing in Publication Data
Tayler, A. B.
Mathematical models in applied mechanics.—
(Oxford applied mathematics and computing
science series; v. 1)
1. Differential equations, Partial
2. Mathematical models
I. Title
515.3'53   QA374
ISBN 0-19-853533-3
ISBN 0-19-853541-4 Pbk

Library of Congress Cataloging in Publication Data
Tayler, Alan B.
Mathematical models in applied mechanics.
(Oxford applied mathematics and computing science
series)
Bibliography: p.
Includes index.
1. Differential equations, Partial—Numerical
solutions. 2. Mechanics, Applied—Mathematical models.
I. Title.   II. Series.
QA374.T378   1985   515.3'53   85-22398
ISBN 0-19-853533-3
ISBN 0-19-853541-4 (pbk.)

Typeset and printed by The Universities Press (Belfast) Ltd

Preface

Since 1975 the graduate M.Sc. programme in Applied Mathematics at Oxford University has contained a full-year course which develops the theory and applications of partial differential equations through the solution of a series of appropriate problems. Most of the problems, which may vary from year to year, have arisen from industrial research through the Oxford Study Groups with Industry, and particular attention is paid to the formulation of each problem. This book is based on the experience of the author and his colleagues in teaching this course and in developing the Study Group project. Its objectives are to demonstrate the skills of formulating problems in Applied Mechanics at the same time as coherently developing the theory and methods of partial differential equations.

The book is not intended as a research monograph on industrial problems nor as a standard text on differential equations, which, however, are its main mathematical theme. Rather, it is an attempt to demonstrate the power of mathematics and the diversity of its application, in the hope of stimulating readers with some background knowledge of Applied Analysis and Mechanics into constructing their own models of new situations. It is designed to be a coherent text so that each problem described develops the theme, and each extension of the theme leads on to a new problem. To achieve this, compromises have had to be made both in the requirements for background mathematical knowledge of the reader and in the extent to which the solution of each problem has been developed. These compromises, and the selection of suitable problems, have been the subject of much discussion for a period of about ten years and I must acknowledge the immense contribution of my colleagues John and Hilary Ockendon.

A further compromise, necessary to avoid excessive length, is that all discussion of numerical methods has been omitted. A problem is considered to be solved if it is in a form for which a well-established numerical method will quickly produce good answers. A number of the problems have, however, stimulated new numerical methods and where this is so a reference will be given to an appropriate description of the method. The approach and style adopted in presenting material is in the same spirit as that of *Mathematics applied to deterministic problems in the natural sciences* by Lin and Segal (1974), but this book has a less broad range and requires a higher level of background knowledge from the reader.

There appears to be no obvious way of teaching mathematical modelling, and experience is clearly of considerable importance. Students can gain experience from the organized case history approach provided the problems are sufficiently novel, and successful models do provide valuable analogies for new situations. It is my hope that this book will provide for the student a useful approach to modelling, for the applied mathematician some diverse applications of the theory of partial differential equations, and for the pure mathematician some questions to answer.

Oxford   A.B.T.
December 1984

Contents

1 Problem formulation

1.1. Introduction

Applied Mechanics is mathematics applied to those problem areas of science which arise from continuum considerations. Thus a macroscopic rather than an atomic or quantum view of the phenomena is adopted and attention directed to those areas in which observations or experiments are reproducible, and measured data are available in some form. The objective is to construct a mathematical simulation, or model, of a given scientific phenomenon, which agrees with existing measurements to within a specified accuracy and can be used with confidence to predict future observations and behaviour. Stemming from this continuous viewpoint the mathematical models will naturally involve relationships between continuous functions of space and time which describe the application of the fundamental scientific principles of homogeneity, isotropy, and conservation to a given problem area. (The simplest and perhaps most celebrated example is Newton's Second Law of Motion for a particle, Force = Mass × Acceleration, which leads to the model for a particle falling under gravity of $d^2s/dt^2 = -g$ and predicts that $s = -\frac{1}{2}gt^2 + ut$.) Hence, to pursue Applied Mechanics successfully two contrasting but complementary skills are needed, namely the ability to formulate a given problem in appropriate mathematical terms, and sufficient knowledge to obtain useful information from that mathematical model. The skill in formulation lies in finding a model which is simple enough to give useful information easily, but which is diverse enough to give all the information required with sufficient accuracy. A model might be appropriate in one set of circumstances but of no value in others because it is either over-complicated or too simple. Once a model is well established it is rarely presented in a textbook in such a way that the reader can appreciate both the repeated modifications made to earlier less satisfactory models, and the interplay between these earlier models, their analysis, and the comparison of prediction with experiment. The use of analogy between well-established models and new problems is clearly a valuable aid, and in the next two sections of this chapter a brief account of the various stages in the formulation of a mathematical model will be illustrated by models for aspects of fluid mechanics and heat transfer. Emphasis will be placed on the modelling aspects of these well-established topics and reference given to other texts for their detailed analysis and development.

1.2. Normalization of the problem

A macroscopic problem will in general have a number of quantitative physical features, varying in space and time, to which mathematical symbols may be assigned. The features may be represented by scalar, vector or tensor quantities and will be functions of space and time variables which are continuous and differentiable except on special surfaces. These quantities will be related by the fundamental physical laws and such relationships will introduce a number of measurable physical constants. Any observed phenomenon will depend on the physical features and constants, but in a given situation it may not be clear which are the important ones. Thus, in attempting to construct a model, judgement has to be made about which features to include and which to neglect. If an important feature is omitted then the model will not describe the observed phenomenon accurately enough or the model may not be self-consistent. If unnecessary features are included then the model will be more difficult to solve because of its increased complexity. Thus it is advisable to adopt a simple approach with a minimum of features included at the first attempt, and additional features added, if necessary, one by one. It should then be possible to estimate the effect of the additional feature on the results of the original model. This is done by *normalizing* the problem, that is defining it in terms of non-dimensional variables whose typical scale is of order one, and the relative magnitude of different physical effects is measured by non-dimensional parameters or dimensionless groups. A good example is that of fluid mechanics where the Navier–Stokes equations for the flow of an incompressible viscous fluid provide a well-established model. The unknown quantities are the velocity q, pressure p, and temperature T as functions of space variables x and time t, where the vectors have three components. Application of the physical laws of conservation of mass, momentum, and energy as in, for example, Landau and Lifshitz (1963) leads to the following model:

$$\nabla \cdot q = 0, \tag{1.1}$$

$$\rho\left[\frac{\partial q}{\partial t} + (q \cdot \nabla)q\right] = -\nabla p + \mu \nabla^2 q + \rho F, \tag{1.2}$$

$$\rho c\left[\frac{\partial T}{\partial t} + (q \cdot \nabla)T\right] = k\nabla^2 T + \mu \sum_{i=1}^{3} \sum_{j=1}^{3} \left(\frac{\partial q_i}{\partial x_j}\right)^2. \tag{1.3}$$

The constants involved are the density ρ, the coefficient of viscosity μ, the specific heat c, the conductivity k, and if the body force F is due to gravity, the gravitational acceleration g. In eqn (1.2) the physical effects modelled are inertia, viscous forces, and gravity; in eqn (1.3) they are heat convection (sometimes called advection), heat conduction, and viscous dissipation. In writing down this model a judgement has already been made that variations in density and viscosity coefficient may be neglected, and that the stress in the fluid depends linearly on the rate of strain. To determine whether more physical effects may be neglected, typical scales or reference values must be chosen for each variable so that the problem can be normalized. This requires that the problem be completed by the addition of boundary conditions. Suitable boundary conditions should be such that they lead to a well-posed boundary value problem (that is, one with a unique solution) for at least some range of values of the constants involved, and this requirement will feature strongly in many of the problems to be discussed. Assuming suitable conditions are given they will contain reference or scale values for the variables, say U for q, L for x, L/U for t, and $T_1 - T_0$ for T, where T_0 is the ambient temperature. If we then define non-dimensional variables by

$$\bar{x} = \frac{x}{L}, \quad \bar{q} = \frac{q}{U}, \quad \bar{t} = \frac{tU}{L}, \quad \bar{p} = \frac{p}{\rho U^2}, \quad \bar{T} = \frac{T}{T_1 - T_0}$$

the model eqns (1.1)–(1.3) become

$$\nabla \cdot \bar{q} = 0,$$

$$\frac{\partial \bar{q}}{\partial \bar{t}} + (\bar{q} \cdot \nabla)\bar{q} = -\nabla \bar{p} + \frac{1}{Re} \nabla^2 \bar{q} - \frac{1}{Fr} k,$$

$$Re\left(\frac{\partial \bar{T}}{\partial t} + (\bar{q} \cdot \nabla)\bar{T}\right) = \frac{1}{Pr} \nabla^2 \bar{T} + \frac{Br}{Pr} \sum_{i=1}^{3} \sum_{j=1}^{3} \left(\frac{\partial \bar{q}_i}{\partial \bar{x}_j}\right)^2,$$

where the operator ∇ is with respect to $\bar{x}$ variables and k is a unit vector in the upward vertical direction. Four non-dimensional parameters are needed, although clearly different combinations are possible. We have chosen the *Reynolds* number $Re = \rho UL/\mu$, which compares the effects of inertia and viscosity, the *Froude* number $Fr = U^2/Lg$, which compares inertia and gravity, the *Prandtl* number $Pr = \mu c/k$, which compares the viscous time scale with that of heat conduction, and the *Brinkman* number $Br = \mu U^2/[k(T_1 - T_0)]$, which compares viscous dissipation of heat with heat conduction. A further parameter, which is commonly used, compares the inertial time-scale with that of conduction and is the *Peclet* number $Pe = RePr$.

The numerical values of these numbers must be established for any given problem, and one or more of them may be very small or large. Thus for large Froude number the effects of gravity may be neglected since they only appear in a single term with a small coefficient; for small Brinkman number, dissipation may be neglected in comparison with heat conduction and a simpler model derived. It may, however, be necessary, if there is no obvious choice of scale, to rescale one or more of the variables before neglecting terms multiplied by small parameters. Thus if $Re \ll 1$, in addition to $1/Fr$ and $Br \ll 1$, the pressure term will disappear on putting $Re = 0$ and it will not be possible to satisfy all the boundary conditions because the order of the system of equations has been reduced. The reason is that the pressure scale was based on inertia rather than on viscous forces, and a rescaling $\bar{p} = \bar{p}Re$ is necessary. In this case, with $Re = 0 = Br = 1/Fr$, and Pr not large so that $Pe = 0$, the model reduces to

$$\nabla \bar{q} = 0 = \nabla^2 \bar{T}, \quad \nabla^2 \bar{q} = \nabla \bar{p}. \tag{1.4}$$

However, eqn (1.4) does not contain any time derivatives, that is it describes an instantaneous situation on the inertia time-scale.

A viscous time-scale would be $\rho L^2/\mu$† and with $\bar{\bar{t}} = t/(\rho L^2/\mu) = \bar{t}/Re$, the model becomes

$$\frac{\partial \bar{\bar{q}}}{\partial \bar{\bar{t}}} = \nabla^2 \bar{q} - \nabla \bar{\bar{p}}, \tag{1.5}$$

$$\frac{\partial \bar{T}}{\partial \bar{\bar{t}}} = \frac{1}{Pe} \nabla^2 \bar{T}. \tag{1.6}$$

It may also be the case that the temperature time-scale t_0 is neither the inertia nor the viscous time-scale and eqn (1.6) should be replaced by $\partial \bar{T}/\partial \bar{t} = \bar{\kappa}^2 \nabla^2 \bar{T}$, where $\bar{\kappa}^2 = k t_0/\rho c L^2$ is the non-dimensional *diffusivity*.† A different balance of terms, when Br is not small, is discussed in Exercise 1.1.

1.3. Perturbation procedures

The simplification of the model obtained by neglecting a term multiplied by a small non-dimensional parameter can be thought of as the first step in a systematic procedure for obtaining an asymptotic expansion for the full solution in terms of the small parameter. Thus the solution for the pressure $\bar{p}$ (and also $\bar{T}$ and $\bar{q}$) in the example of the previous section (with $1/Fr = 0 = Br$) has the functional form $\bar{p} = \bar{p}(\bar{x}, \bar{t}, Re)$, and if it possesses an asymptotic expansion in powers of Re, then

$$\bar{p} = \bar{p}_0(\bar{x}, \bar{t}) + Re \; \bar{p}_1(\bar{x}, \bar{t}) + \dots$$

for sufficiently small values of Re. The solution to the simplified model in which $Re = 0$ is then $\bar{p}_0$, and higher order terms can be successively obtained. This is a *regular perturbation* procedure and may be used if $\bar{p}(\bar{x}, \bar{t}, Re)$ (and also $\bar{T}$ and $\bar{q}$) are regular functions of Re near $Re = 0$. However this is almost always not known in advance and the approach adopted is to attempt to perform a regular expansion and expect that if it is not valid there will be a clear inconsistency in the first few terms of the expansion. This inconsistency may arise in the form of the problem for $\bar{p}_0$ being ill posed, as happened before $\bar{p}$ was rescaled when there was no solution for $\bar{p}_0$. Alternatively the solutions obtained may not be bounded in some region of the space and time variables $\bar{x}$

---

† The dimensions of the kinematic viscosity $\nu = \mu/\rho$, and the dimensional diffusivity $k/\rho c$, are (length)^2/time.

and $\bar{t}$. In such a situation the expansion is singular, and a *singular perturbation* procedure must be adopted. An example is the problem of the previous section when $Re \gg 1$ so that a regular expansion would be in powers of $1/Re$. When $1/Re$ is put equal to zero the order of the equation is reduced because the highest derivative term disappears and not all the boundary conditions can be satisfied. An alternative statement is that, adjacent to a boundary on which q is given, some derivatives of q will be unbounded and a boundary layer is formed in which derivatives of q are large and the neglected term must be retained. A detailed discussion of such problems involving the technique of matched expansions will be given in Chapter 5. In Chapters 1 to 4 the models will all be obtained by regular expansions.

Problem A. Furnace reaction analysis

One technique for analysing the properties of a given chemical sample is to heat it continuously in a furnace to a temperature at which a reaction takes place, and to compare the temperature profile in the sample with that of the furnace. Because of the scale of the furnace, heat conduction in the sample will be important and the heat conductivity k and diffusivity κ^2 are assumed to be known constants. The problem is to determine the constants which describe the reaction, namely the heat of reaction λ, the activation energy E, and the rate constant A^*, from these observations. For simplicity, we consider only a single first-order reaction with one rate constant, so that if $\alpha(t)$ is the mass fraction of active material, then from the Arrhenius law

$$\frac{\mathrm{d}\alpha}{\mathrm{d}t} = -A^*\alpha \mathrm{e}^{-E/RT}, \tag{A1}$$

where R is a known constant and T is the absolute temperature. Also, for simplicity, we consider only a one-dimensional model, that is the sample is an infinite block of width $2L$, heated from both sides with no heat transfer parallel to its sides. Then α is a function of x and t and the time derivative in eqn (A1) is the partial derivative $\partial\alpha/\partial t$. The chemical reaction, when it occurs, creates heat in the sample so that an energy or heat balance, similar to eqn (1.3) but with $q = 0$ (because it is a solid), gives

$$k\frac{\partial^2 T}{\partial x^2} = \rho c \frac{\partial T}{\partial t} + \rho\lambda\frac{\partial\alpha}{\partial t}. \tag{A2}$$

The boundary conditions are that no reaction has taken place at $t = 0$ so that $\alpha = 1$, and $\alpha \geqslant 0$ for all t. Also $T = T_0$ at $t = 0$ and, if we assume a symmetrical situation $\partial T/\partial x = 0$ at $x = 0$. For the heating due to the furnace at $x = L$, a simple model is to assume that a heat transfer coefficient h^* from the furnace to the sample is known so that

$$k\frac{\partial T}{\partial x} = -h^*[T - T_F(t)], \tag{A3}$$

where T_F is the prescribed furnace temperature. Again for simplicity we consider a problem in which the furnace is heated at a uniform rate β so that $T_F(t) = T_0 + \beta t$ where β is known and can be varied.

A typical temperature profile which is observed in practice is shown in Fig. A1, and the first task is to determine for which range of values of the constants such a profile might be predicted by the mathematical model. To do this we first normalize the problem, scaling the variables x with L, t with the conduction time L^2/κ^2, and $T - T_0$ with λ/c. Equations (A2) and (A1) are now expressed using scaled variables, where without ambiguity the same symbol has been used with no superscript bars:

$$\frac{\partial^2 T}{\partial x^2} = \frac{\partial T}{\partial t} + \frac{\partial \alpha}{\partial t}; \quad 0 < x < 1;$$

$$-\frac{\partial \alpha}{\partial t} = \tilde{A}\alpha \exp\left(-\frac{1}{\varepsilon(1 + \delta T)}\right);$$

where $\tilde{A} = A^*L^2/\kappa^2$, $\varepsilon = RT_0/E$, and $\delta = \lambda/cT_0$.

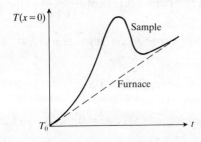

Fig. A1. Temperature profile in a furnace

The boundary conditions are

$$x = 0, \quad \frac{\partial T}{\partial x} = 0; \quad x = 1, \quad \frac{\partial T}{\partial x} = -h(T - \omega t); \quad t = 0, \quad \alpha = 1, \quad T = 0;$$

where $h = h^* L/k$ and $\omega = \beta(L^2 c/\kappa^2 \lambda)$.

There are five non-dimensional parameters, three of which are unknown constants, h (which is given), and ω which is not known but can be varied. If all of them are of order one the problem is clearly difficult and little analytical progress can be made. We therefore consider the implications of a situation in which $\omega \ll 1$, that is the furnace is only slowly heated compared with the conduction time. For the furnace heating term to appear in the equations in the limit $\omega \to 0$ time has to be rescaled, so we finally write $\tau = \omega t$ to obtain the boundary value problem

$$\frac{\partial^2 T}{\partial x^2} = \omega\left(\frac{\partial \alpha}{\partial \tau} + \frac{\partial T}{\partial \tau}\right), \tag{A4}$$

$$-\frac{\partial \alpha}{\partial \tau} = A\alpha \exp\left(-\frac{1}{\varepsilon(1 + \delta T)}\right), \tag{A5}$$

where $x = 0$, $\partial T/\partial x = 0$; $x = 1$, $\partial T/\partial x = -h(T - \tau)$; $\tau = 0$, $\alpha = 1$, $T = 0$; and $A = \tilde{A}/\omega$.

We now look for a regular expansion of T and α in powers of ω, and to retain a non-trivial relation for α we require that A is of order one, and hence $\tilde{A}$ is small. A simple integration gives

$$T = \tau + \omega T_1(x, \tau) + O(\omega^2),$$

and hence $\alpha = \alpha_0(\tau) + O(\omega)$ where

$$\alpha_0(\tau) = \exp\left[-A \int_0^\tau \exp\left(\frac{-1}{\varepsilon(1 + \delta \tau')}\right) d\tau'\right]. \tag{A6}$$

Substituting these expressions into eqn (A4) and neglecting $O(\omega^2)$:

$$\frac{\partial^2 T_1}{\partial x^2} = 1 + \dot{\alpha}_0,$$

with boundary conditions $\partial T_1/\partial x = 0$, $x = 0$ and $\partial T_1/\partial x = -hT_1$, $x = 1$.

From a further integration

$$T_1 = (1 + \dot{\alpha}_0)\left(\frac{x^2 - 1}{2} - \frac{1}{h}\right). \tag{A7}$$

This gives a consistent expansion for T if the constants involved are all of order one, except near $\tau = 0$ where the boundary condition $T = 0$ has not been satisfied. However, we have satisfied $\alpha = 1$ and could rescale for small τ with $\alpha = 1$ to make good this deficiency; it is not, however, of great interest since at such times the reaction has hardly started. The question of interest is whether eqn (A7) gives a curve for T at $x = 0$ of the general form shown in Fig. A1; that is, can constants A, ε and δ be chosen so that the observed results are fitted by

$$T_{x=0} = \tau + \omega\left(\frac{1}{2} + \frac{1}{h}\right)$$

$$\times \left\{ A \exp\left[\frac{-1}{\varepsilon(1 + \delta\tau)} - A\int_0^\tau \exp\left(\frac{-1}{\varepsilon(1 + \delta\tau')}\right) d\tau'\right] - 1\right\}. \tag{A8}$$

This is a rather formidable expression which may be simplified, and still have a suitable profile, in the case of small activation energy and heat of reaction, that is ε and δ both small (see Exercise 1.2). In general, however, profiles can be constructed for $T_{x=0} - \tau$ which have a maximum value ωT_m at $\tau = \tau_m$. These values are easily measured from the experimental evidence, and provide two relations connecting the unknowns A, ε and δ. A further relation may be obtained by some profile width criterion such as the area under the profile, to complete the determination of A, ε and δ, and hence the constants A^*, E and λ. Varying ω, that is the furnace heating rate β, will give a check on the model and the experimental data.

1.4. Analytical methods for ordinary differential equations

In the following chapters we shall formulate problems as mathematical models which are boundary-value problems involving partial differential equations. To develop the solution of such boundary-value problems requires a knowledge of the solution of ordinary differential equations with initial-value or two-point boundary conditions. There are many texts on this topic with

varied approaches, but for completeness we shall briefly review
three major ideas in the solution of ordinary differential equations.

(a) *The use of transforms.* The Fourier Integral Theorem may be
used to invert the Fourier transform of $f(x)$, defined by

$$F(k) = \int_{-\infty}^{\infty} e^{ikx} f(x) \, dx,$$

by means of the inversion relation

$$f(x) = \frac{1}{2\pi} \int_{-\infty}^{\infty} e^{-ikx} F(k) \, dk, \qquad (1.7)$$

provided both integrals exist.

The value of this inversion relation is that the derivative of $f(x)$
has the transform $-ikF(k)$, so that the operation of differentiating with the x variable is equivalent to multiplying by $-ik$ with
the k variable. Moreover, there is a duality in that the transform
of $xf(x)$ is $-i(dF/dk)$, so that multiplying by ix is equivalent to
differentiating with respect to k. Thus linear constant-coefficient
ordinary differential equations for f with respect to x become
algebraic expressions in k multiplying F, and the solution for F is
trivial. Note that not only must the problem be linear in f, but it
has to be defined on $(-\infty, \infty)$ and f must certainly vanish as
$|x| \rightarrow \infty$. To evaluate the inversion relation of eqn (1.7) the
techniques of complex contour integration are clearly applicable.

The Fourier Integral Theorem may also be used to obtain the
Laplace transform, which is suitable for functions f defined on
$(0, \infty)$ with possibly exponential growth as $x \rightarrow \infty$; it is given by

$$\hat{f}(p) = \int_{0}^{\infty} e^{-px} f(x) \, dx.$$

The inversion relation then becomes

$$f(x) = \frac{1}{2\pi i} \int_{\Gamma} e^{px} \hat{f}(p) \, dp, \quad x > 0,$$

where Γ is a contour in the p plane such that the contour integral
is zero for all $x < 0$. The inversion relation is usually written in

the form

$$f(x) = \frac{1}{2\pi i} \int_{\gamma-i\infty}^{\gamma+i\infty} e^{px} \hat{f}(p) \, dp, \quad x > 0, \tag{1.8}$$

where γ is greater than the real part of any singularity of $\hat{f}(p)$. This transform may be applied to any linear, constant-coefficient, initial-value problem in the form

$$L_n\left(\frac{d}{dx}\right) f = g; \quad f = f' = \ldots = f^{n-1} = 0, \quad \text{at} \quad x = 0,$$

to give $L_n(p)\hat{f} = \hat{g}$, and

$$f(x) = \frac{1}{2\pi i} \int \frac{e^{px} \hat{g}(p) \, dp}{L_n(p)}. \tag{1.9}$$

Here the initial conditions have been taken to be zero, since non-zero conditions can easily be removed by subtracting a suitable polynomial from f and redefining g. Thus the Laplace transform is appropriate for initial-value problems, whereas the Fourier transform is suitable for two-point boundary problems on $(-\infty, \infty)$.

The Fourier transform may be used to derive other transform pairs. A simple example is the Fourier sine transform

$$F(k) = \int_0^\infty f(x)\sin(kx) \, dx$$

obtained by defining $f \equiv 0$ in $x < 0$, real in $x > 0$, and taking the imaginary part of eqn (1.7) to give

$$f(x) = \frac{2}{\pi} \int_0^\infty F(k)\sin(kx) \, dk. \tag{1.10}$$

This is suitable for functions f defined on $(0, \infty)$ which vanish as $x \to \infty$. It is also useful for linear constant-coefficient equations which contain only even derivatives of f, with the condition that even derivatives of f are prescribed on $x = 0$ (f is included as an even derivative). Likewise there is a cosine transform for odd derivatives only.

It is also possible to construct transform pairs for use with certain linear differential equations which do not have constant coefficients. A simple example is Bessel's equation, which frequently occurs in problems with cylindrical symmetry and has the form

$$\frac{d^2f}{dr^2}+\frac{1}{r}\frac{df}{dr}+\left(1-\frac{n^2}{r^2}\right)f = g, \quad n = 0, 1, 2, \dots. \quad (1.11)$$

This equation (with $g \equiv 0$) is obtained from the reduction of $\nabla^2\phi = 0$ into cylindrical polar form $\phi = e^{in\theta}f(r)$, and the solution bounded at $r = 0$ is $J_n(r)$. By applying the Fourier transform result to ϕ for both x and y Cartesian variables, we obtain the double transform result

$$\Phi(k, l) = \int_{-\infty}^{\infty}\int_{-\infty}^{\infty} e^{ikx+ily}\phi(x, y)\,dx\,dy,$$

$$\phi(x, y) = \frac{1}{4\pi^2}\int_{-\infty}^{\infty} e^{-ikx-ily}\Phi(k, l)\,dk\,dl.$$

If $\phi = f(r)e^{in\theta}$ and $\Phi = F(\rho)e^{in\alpha}$ then after some manipulation (see, for example, Mackie (1965) for details) we obtain the Hankel transform pair

$$F(\rho) = \int_0^{\infty} rJ_n(\rho r)f(r)\,dr,$$

$$f(r) = \int_0^{\infty} \rho J_n(\rho r)F(\rho)\,d\rho, \quad (1.12)$$

provided again that both integrals exist. It is easily verified that this is appropriate to use with Bessel's equation, as in Exercise 1.4.

Further transform pairs for other common differential equations may be constructed, and the reader is referred to a textbook on transforms such as Sneddon (1951) for details, and to Exercises 1.3 and 1.4. However, using transform methods to reduce a problem to an integral, possibly complex, as in eqn (1.9), may not necessarily give a useful representation of the solution if the properties of that integral are difficult to evaluate. For both Fourier and Laplace transforms the Convolution Theorem provides one valuable way of dealing with complicated expressions for $\hat{g}$.

For the Fourier transform the theorem states that the transform of the convolution of two functions g and h defined by

$$g * h = \int_{-\infty}^{\infty} g(s)h(x-s)\,ds$$

is the product GH of their transforms. Thus if we are solving $L_n(d/dx)f = g$ on $(-\infty, \infty)$, with f vanishing as $|x| \to \infty$, the Fourier transform of f is given by

$$F(k) = \frac{G(k)}{L_n(-ik)}.$$

If we write $H(k) = [L_n(-ik)]^{-1}$, then

$$f(x) = \int_{-\infty}^{\infty} g(s)h(x-s)\, ds,$$

where $h(x)$ is the inverse transform of H, and is simple to evaluate from eqn (1.7) by the calculus of residues. Note that $h(x)$ satisfies $L_n(d/dx)h = \delta(x)$, where δ is the delta function, defined as zero for $x \neq 0$ but $\int \delta(x)\, dx = 1$ for any interval including $x = 0$. An alternative formulation, changing the variable from x to $x - s$, is

$$f = \int_{-\infty}^{\infty} g(s)G(x, s)\, ds, \qquad (1.13)$$

where

$$L_n\left(\frac{d}{dx}\right)G = \delta(x-s) \qquad (1.14)$$

and G vanishes as $|x| \to \infty$.

(b) *The use of Green's functions.* Equation (1.13) suggests that we may be able to express the solution of any linear two-point boundary problem in the form

$$f = \int_{a}^{b} g(s)G(x, s)\, ds, \qquad (1.15)$$

where G depends only on the differential operator L_n. For simplicity we discuss this in the case $n = 2$ and consider

$$\frac{d^2f}{dx^2} + c_1(x)\frac{df}{dx} + c_2(x)f = g(x), \qquad (1.16)$$

subject to the homogeneous boundary conditions

$$f + \lambda\frac{df}{dx} = 0 \quad \text{at} \quad x = a, \quad f + \mu\frac{df}{dx} = 0 \quad \text{at} \quad x = b.$$

Substituting eqn (1.15) into eqn (1.16) and inserting boundary conditions, assuming that G is continuous at $x = s$, the equations are satisfied if

$$G + \lambda \frac{\partial G}{\partial x} = 0 \quad \text{at} \quad x = a, \quad G + \mu \frac{\partial G}{\partial x} = 0 \quad \text{at} \quad x = b,$$

(1.17)

and

$$\frac{d}{dx} \int_a^b g \frac{\partial G}{\partial x} ds + \int_a^b g \left[c_1(x) \frac{\partial G}{\partial x} + c_2(x)G \right] ds = g(x) \quad \text{for} \quad a < x < b.$$

If G is discontinuous at $x = s$, but G satisfies

$$\frac{\partial^2 G}{\partial x^2} + c_1(x) \frac{\partial G}{\partial x} + c_2(x)G = 0, \quad a < x \neq s < b, \quad (1.18)$$

then the necessary condition is

$$\left[g \frac{\partial G}{\partial x} \right]_{x+0}^{x-0} = g(x).$$

This is satisfied by requiring that the discontinuity in $\partial G/\partial x$ is 1. The continuity conditions on G and $\partial G/\partial x$, together with eqns (1.17) and (1.18), complete the definition of the *Green's function*, G. An equivalent statement is eqn (1.17) together with

$$\frac{\partial^2 G}{\partial x^2} + c_1(x) \frac{\partial G}{\partial x} + c_2(x)G = \delta(x - s),$$

since integrating this once with respect to x gives

$$\left[\frac{\partial G}{\partial x} \right]_{x=s-0}^{x=s+0} = 1.$$

Exercise 1.5 illustrates the determination of G. The appropriate definition of G for $n > 2$ is that it satisfies the homogeneous boundary conditions and eqn (1.14), where L_n need not have constant coefficients. Inhomogeneous boundary conditions must first be absorbed into the definition of g by subtraction of a suitable polynomial. It may be difficult to find a representation for G in terms of simple functions, but it can be shown to exist and to be unique under suitable conditions, and it is defined independently of g.

A particularly valuable use of the Green's function is to convert an ordinary differential equation into an integral equation. Thus for eqn (1.15) the Green's function G_0 for the case $c_2 \equiv 0$ can always be found in terms of quadratures, and may be used when $c_2 \neq 0$ to give

$$f(x) = \int_a^b [g(s) - c_2(x)f(x)]G_0(x, s)\, ds,$$

which is an integral equation for f. Any linear second-order two-point boundary problem can therefore be reduced to a linear integral equation of the general form

$$\lambda f(x) = \int_0^1 K(x, s)f(s)\, ds + k(x), \qquad (1.19)$$

where λ is a parameter introduced for convenience and the range has been scaled to $(0, 1)$. If this integral equation has its *kernel* K symmetric, that is $K(x, s) = K(s, x)$, and is derived from a differential equation, then that equation is said to be *self-adjoint*. With $\lambda \neq 0$ eqn (1.19) is a Fredholm integral equation of the second kind, and an important feature is that it has a finite range.

Although it is no easier to find solutions in terms of simple functions for the integral equation than for the differential equation, the integral formulation is the starting point for proofs of existence and of uniqueness of solutions, and for convergent numerical procedures. The most important result is that for a large class of kernels K there will exist special values of λ, called *eigenvalues* $\lambda_r, r = 1, \ldots, n$ where n may be unbounded, for which there exist non-zero solutions $f = f_r$, called *eigenfunctions*, of the homogeneous problem $k \equiv 0$. Moreover, if K is symmetric these eigenfunctions are orthogonal, that is

$$\int_0^1 f_r(s)f_q(s)\, ds = 0$$

if $r \neq q$. If λ is not an eigenvalue then there exists a unique solution of eqn (1.19).

However, if λ is an eigenvalue λ_r then the *Fredholm alternative* applies, and (for K symmetric) there is no solution given that

$$\int_0^1 k(s)f_r(s)\, ds \neq 0;$$

if however

$$\int_0^1 k(s)f_r(s)\,\mathrm{d}s = 0$$

then a solution exists but is only determined up to an additive term $a_r f_r(x)$ (see Exercise 1.6). In this simple statement of the Fredholm theory we have assumed that for each eigenvalue there is only one independent eigensolution; the general case with multiple eigensolutions and non-symmetric kernels has a similar form, and details are given in Chapter 3 of Courant and Hilbert (1976). If the integral equation had been derived from a differential equation (which will now contain the parameter λ), then it too will possess eigenvalues for which there exist eigenfunctions as solutions to the homogeneous problem.

A further extension of the theory shows that for a self-adjoint two-point boundary problem the solution when λ is not an eigenvalue may be expressed as a linear combination of all the eigenfunctions. In general there will be an infinite number of eigenvalues and eigensolutions, and this infinite linear combination will be convergent. The well-known example is that of a Fourier series when $\mathrm{d}^2f/\mathrm{d}x^2 + \lambda f = 0$, with (for example) conditions $f(0) = f(\pi) = 0$. In this case $\lambda = r^2$, $r = 0, 1, 2\ldots$, and $f_r = \sin rx$ so that

$$f = \sum_1^\infty a_r \sin rx,$$

the Fourier half-range sine series. The coefficients a_r may be determined from the orthogonality property of eigenfunctions, and the use of such orthogonal expansions is common in applied mathematics. However, like the integral representation by transforms, it may not necessarily be the best representation of the solution for obtaining numerical results to the two point boundary problem.

(c) *Asymptotic methods.* In looking for solutions of ordinary differential equations we must be clear what exactly is required by a 'solution'. In many cases a qualitative sketch or graph of the function f for all values of the independent variable x may suffice, or at least be of primary importance, with detailed numerical values being unnecessary. In such cases attention is focused

on singular points of the equation, where the solution is unbounded or not differentiable, and the nature of these singular
points may be examined by approximations which are valid in
their neighbourhood. To be precise about the validity of the
approximation the ideas of asymptotic analysis are needed, and a
brief review will be given below. For a detailed exposition see
Bender and Orszag (1978).

First we introduce the order symbols O and o, where $f(x) =$
$O[g(x)]$ as $x \to x_0$ implies (for $g(x) \neq 0$) that $|f/g| < M$ as $x \to x_0$,
and we say that f is of the same order of magnitude as g. If
$f(x) = o[g(x)]$, then $|f/g| \to 0$ as $x \to x_0$ and f is of smaller order
of magnitude than g. The simpler notation $f \sim g$ implies that
$f = g + o(g)$ and gives more information than $f = O(g)$. Thus in
this sense g is an approximation to f and $o(g)$ replaces the crude
statement 'plus smaller terms'.

An asymptotic sequence $g_r(x)$, $r = 0, 1, 2, \ldots$ is defined by
$g_{r+1} = o(g_r)$ for all r as $x \to x_0$, and a special case is $g_r = (x - x_0)^r$.
Usually the singular point is taken to be at $x = 0$ or $x \to \infty$, and in
the latter case an asymptotic sequence is $g_r = 1/x^r$. An asymptotic
expansion of f as $x \to 0$ is given by

$$f(x) = \sum_0^n a_r g_r(x) + o(g_n), \qquad (1.20)$$

which is not unique unless the sequence $g_r(x)$ is prescribed. An
asymptotic series is

$$f(x) = \sum_0^n a_r x^r + o(x^n),$$

which need not be a convergent power series since the test for
convergence is that f tends to a limit as $n \to \infty$ for fixed $x > 0$,
whereas for an asymptotic series n is fixed and $x \to 0$.

In developing an asymptotic expansion of a function there are
two features; to recognize an appropriate asymptotic sequence g_r
and then to find coefficients a_r such that the remainder $f - \sum_0^n a_r g_r$
may be proved to be $o(g_n)$. To illustrate these features consider
evaluating asymptotically the integral

$$f(x) = \int_0^x e^{-t} t^{a-1} \, dt, \quad a > 0; \quad x > 0.$$

As $x \to 0$ we can obtain a plausible asymptotic expansion for f by integrating by parts, so that

$$f(x) = \left[\frac{e^{-t}t^a}{a}\right]_0^x + \frac{1}{a}\int_0^x e^{-t}t^a \, dt$$

$$= \frac{e^{-x}x^a}{a} + \frac{e^{-x}x^{a+1}}{a(a+1)} + \ldots + \frac{a!}{(a+n)!}\left(e^{-x}x^{a+n} + \int_0^x e^{-t}t^{a+n} \, dt\right).$$

The integral remainder term is easily shown to be $o(e^{-x}x^{a+n})$ as $x \to 0$ so that

$$e^{-x}x^a \sum_0^n \frac{x^r}{(a+r)\ldots a}$$

is an asymptotic expansion for f. It is not an asymptotic series, but could be rewritten as a series when a is an integer by expanding e^{-x} in a power series up to terms x^n, and multiplying the two expressions together.

As $x \to \infty$ we also integrate by parts, but first subtract out a constant. In this case

$$f(x) = \int_0^\infty e^{-t}t^{a-1} \, dt - \int_x^\infty e^{-t}t^{a-1} \, dt$$

$$= \Gamma(a) - [-e^{-t}t^{a-1}]_x^\infty - (a-1)\int_x^\infty e^{-t}t^{a-2} \, dt,$$

where Γ is the gamma function. The asymptotic expansion obtained by this procedure is

$$f(x) \sim \Gamma(a) - e^{-x}x^a\left(\frac{1}{x} + \frac{a-1}{x^2} \ldots \frac{(a-1)\ldots(a-n+1)}{x^n}\right),$$

where again the integral remainder is easily shown to be $o(e^{-x}x^{a-n})$. In this case it is not possible to rewrite the expression as an asymptotic series in $1/x$ because e^{-x} is smaller than any power of x as $x \to \infty$.

Some examples may require considerable ingenuity to construct an appropriate asymptotic expansion, both in the choice of sequence and in the proof that the remainder is small enough (see for example Exercise 1.7). A class of integrals which give rise to asymptotic expansions that cannot be represented in a series form as $x \to \infty$ is given by

$$f(x) = \int_0^1 e^{xh(t)} g(t) \, dt, \quad x > 0,$$

where h and g are real regular functions on $(0, 1)$.

There are two general cases to consider:
(i) $h(t)$ *has a maximum value at an end point, say* $t = 0$ *and* $h'(0) < 0$; *and*
(ii) $h(t)$ *has a maximum at an interior point* $t = t_0$, *so that* $h'(t_0) = 0$ *and* $h''(t_0) < 0$.
A heuristic argument about the size of $e^{xh(t)}$ suggests that to obtain an appropriate asymptotic sequence the integrand should be expanded about $t = 0$ in case (i), and about $t = t_0$ in case (ii), since the neighbourhood of these values of t 'dominates' the integral. With a little manipulation this procedure suggests the sequences $e^{xh(0)}/x^{r+1}$ for case (i) and $e^{xh(t_0)}/x^{r+\frac{1}{2}}$ for case (ii). The actual proof that expansion in these sequences is asymptotic is not straightforward, but it is based on the proof for the case $h(t) \equiv -t$, called Watson's Lemma; for details see Murray (1984). If $g(t)$ is singular at $t = 0$ in case (i) or $t = t_0$ in case (ii), or if $h'(0) = 0$ in case (i) or $h''(t_0) = 0$ in case (ii), then the appropriate asymptotic sequences are modified.

A further extension is to consider the class of integrals

$$f(x) = \int_0^1 e^{ixh(t)} g(t) \, dt,$$

where again h and g are real regular functions on $(0, 1)$. A heuristic argument suggests that the dominant contribution to the integral comes from the neighbourhood of $t = t_0$, where $h'(t_0) = 0$. Expanding about $t = t_0$ with $h''(t_0) \neq 0$ we are led to an asymptotic sequence $e^{ixh(t_0)}/x^{r+\frac{1}{2}}$, but the proof that the corresponding expansion is asymptotic is again lengthy. The use of only one term in the expansion $(r = 0)$ as an asymptotic representation for f is called the *method of stationary phase*.

Asymptotic methods may be applied directly to differential equations, and most textbooks describe the method of solution in series. This is applied at a regular singular point of the equation where an expansion of the form $\sum_0^\infty a_r x^{r+c}$ is constructed, identically satisfying the equation and hence valid as a solution within its radius of convergence. If it is truncated at $r = n$ this will be an asymptotic expansion as $x \to 0$, since the remainder will certainly be $o(x^n)$ if the radius of convergence is not zero. The method also includes the possibility of expansions in the form $\log x \sum_{0}^{\infty} a_r x^r$ which when truncated is again an asymptotic expansion. For an irregular singular point the method of solution in series fails, because the appropriate asymptotic sequences are no longer expressible as non-integral powers (or integral powers multiplied by $\log x$). Such points are usually banished to infinity, and we describe below an appropriate asymptotic expansion for linear second-order differential equations which have an irregular singular point as $x \to \infty$. Consider

$$f'' + h(x)f = 0, \quad \text{where} \quad h(x) \gg O\left(\frac{1}{x^2}\right) \quad \text{as} \quad x \to \infty.$$

Any linear second-order equation can be transformed into this normal form, but a series expansion in $1/x$ is not possible. We look for an asymptotic solution in the form

$$f = \exp(g_0 + g_1 + \ldots + g_n),$$

where g_r is an asymptotic sequence. Substituting in the equation we obtain from each order of magnitude the relations

$$g_0'^2 + h = 0,$$
$$g_0'' + 2g_0'g_1' = 0,$$
$$g_1'' + g_1'^2 + 2g_0'g_2' = 0, \text{ etc.}$$

Using only two terms g_0 and g_1 we obtain the asymptotic expression

$$f \sim \frac{\exp\left\{\int\limits^{x} [-h(t)]^{\frac{1}{2}} \, dt\right\}}{[-h(x)]^{\frac{1}{4}}}, \tag{1.21}$$

and this representation is called the *WKB method* of solution, after Wentzel, Kramer, and Brillouin, who pioneered its use in quantum mechanics. The proof that it is asymptotic as $x \to \infty$ is difficult, and requires that $h(x) \neq 0$ for large enough x. It may be possible to further simplify eqn (1.21) by expanding h asymptotically as $x \to \infty$.

Bessel's equation (1.11) can be reduced to normal form by the transformation

$$u = x^{\frac{1}{2}} f$$

and

$$u'' + [1 + (1 - 4n^2)/4x^2] u = 0.$$

In this case $h \sim 1$, $u \sim C e^{\pm ix}$ and $f \sim C[(e^{\pm ix})/x^{\frac{1}{2}}]$, where C is an arbitrary constant. The disappointing feature of this result is that it only tells us that the Bessel function $J_n(x)$ (the solution which is bounded as $x \to 0$) has the asymptotic value $(\alpha \sin x + \beta \cos x)/x^{\frac{1}{2}}$ as $x \to \infty$, and gives no information about the values for the coefficients α and β. For the Bessel function of imaginary argument, $J_n(ix)$, this is more dramatic since the two asymptotic forms $e^{\pm ix}/x^{\frac{1}{2}}$ have very different behaviour as $x \to \infty$. To obtain a more precise result we first need an integral representation for J_n obtained by taking transforms as in exercise 1.3a.

Further asymptotic methods involving the solution of ordinary differential equations containing a small parameter ε will be discussed at the beginning of Chapter 5.

1.5. Classical analogies

It is of great value in attempting to formulate new problems in mathematical terms to have experience of successful models; moreover many new problems are extensions or variations of the classical models. The problems which will be discussed in the following chapters arise from Applied Mechanics, and the most central and successful classical models are those of Fluid and Solid Mechanics. There are many textbooks on these subjects, with a variety of approaches, and for completeness a necessarily brief review of the well established models will be given in this section.

(a) *Viscous flow.* The classical model for the flow of an incompressible viscous fluid has already been stated in eqns (1.1)–(1.3), which are called the Navier–Stokes equations. Inherent in this model is the Newtonian assumption that the stress tensor $\{\sigma_{ij}\}$ (the jth component of the force per unit area on an element of surface with normal in the ith direction) is a *linear* function of the *rate-of-strain* tensor $\{e_{ij}\}$ where $e_{ij} = \frac{1}{2}[(\partial q_i/\partial x_j)+(\partial q_j/\partial x_i)]$. Thus in deriving eqn (1.2) from a momentum balance in the form

$$\rho \frac{dq_i}{dt} = \rho\left(\frac{\partial q_i}{\partial t} + q_j \frac{\partial q_i}{\partial x_j}\right) = \rho F_i + \frac{\partial \sigma_{ij}}{\partial x_j}, \qquad (1.22)$$

a *constitutive law* has been assumed in the form

$$\sigma_{ij} = -p\,\delta_{ij} + 2\mu e_{ij}. \qquad (1.23)$$

In these equations the dummy suffix notation is used, δ_{ij} is the Kronecker delta, and μ and ρ are constants. The body force F_i is assumed to be given, so that eqns (1.1) and (1.2) constitute four scalar equations for four scalar unknowns q and p. Equation (1.3) is then an additional scalar equation for T. On physical or general mathematical grounds we expect to be able to prescribe that $q = 0$ on fixed spatial boundaries, together with initial data, and obtain a solution at least for a limited time interval. The proof of existence and uniqueness is difficult, and is certainly restricted to a limited time. Many flows become unstable after a finite time, exhibiting the phenomenon of turbulence, and the model is then inadequate; thus stable steady flows may not necessarily exist. As was seen in Section 1.2 the solutions depend on the parameter *Re*, called the Reynolds number, and it can be shown that in a given configuration the steady flow will be stable if the Reynolds number is sufficiently small ($<O(10^3)$), which is in agreement with observation. Thus the model is only appropriate for laminar flows when the Reynolds number is less than its critical value. For a detailed discussion of this extremely complex phenomenon see Landau and Lifshitz (1963).

The Navier–Stokes equations are difficult to solve even for two-dimensional steady flows because of the non-linearity in the convective term $q_i(\partial q_i/\partial x_i)$, or $(q \cdot \nabla)q$ in vector notation. Considerable attention has therefore been paid to simplifications obtained by taking the limits of small and large Reynolds number. The 'slow flow' model ($Re \to 0$) is given by eqns (1.1) and (1.4), and in dimensional variables reduces to

$$\nabla \cdot q = 0; \quad \nabla p = \mu \nabla^2 q; \qquad (1.24)$$

or, eliminating p,

$$\text{div } q = 0 = \text{curl}^3 q.$$

For a two-dimensional situation a stream function $\psi(x, y)$ may be introduced where $\boldsymbol{q} = \text{curl}(0, 0, \psi)$ and the vector equations reduce to the biharmonic equation for ψ, namely

$$\left(\frac{\partial^2}{\partial x^2} + \frac{\partial^2}{\partial y^2}\right)^2 \psi = 0. \qquad (1.25)$$

Appropriate boundary conditions would be to prescribe the velocity, equivalent to prescribing $\nabla \psi$, on a spatial boundary. Prescribing the normal and tangential stress, which is the physical condition needed at a free boundary between two fluids, or rather between a fluid and a constant-pressure region modelling a gas, is less simple and will be discussed later in the context of plane elasticity. The slow-flow model is well posed, and considerable mathematical effort has been expended in constructing solutions, both in terms of simple functions using complex variable methods and by numerical methods.

(b) *Inviscid flows.* The alternative limit is to consider high Reynolds number laminar flows, equivalent to neglecting viscosity; this leads to Euler's equation for an inviscid fluid,

$$\frac{\partial \boldsymbol{q}}{\partial t} + (\boldsymbol{q} \cdot \nabla)\boldsymbol{q} = -\frac{\nabla p}{\rho} + \boldsymbol{F}, \qquad (1.26)$$

where $\boldsymbol{F}$ is any body force per unit mass. Together with $\nabla \cdot \boldsymbol{q} = 0$ these are four scalar equations for four unknowns, but the order of the equation has decreased due to the neglect of the $\nabla^2 \boldsymbol{q}$ term in eqn (1.2). Thus it is not surprising that the velocity can no longer be specified on a fixed boundary. Physically, with no viscosity, slip is allowed and we cannot expect to specify the tangential velocity at a boundary. Thus an appropriate boundary condition on a fixed boundary is that q_n, the normal component of $\boldsymbol{q}$, is equal to zero; this, together with an initial condition, completes the inviscid model. Because of the non-linearity, general solutions are not easy to construct; but it is possible to derive the remarkable result that if the vorticity $\zeta = \text{curl } \boldsymbol{q}$ is initially zero in a given region, then it will remain zero for the same fluid particles for all time. Thus all flows starting from rest or constant conditions are *irrotational*, that is they have curl $\boldsymbol{q} = 0$. This leads to the introduction of a velocity potential ϕ such that $\boldsymbol{q} = \text{grad } \phi$

where ϕ satisfies

$$\nabla^2 \phi = 0, \tag{1.27}$$

and

$$\frac{p}{\rho} + \frac{\partial \phi}{\partial t} + \tfrac{1}{2}(\nabla\phi)^2 = G(t), \tag{1.28}$$

where G is to be determined from an initial condition. The non-linearity now appears only in Bernoulli's equation (1.28), and if $q_n = \partial\phi/\partial n$ is specified on spatial boundaries the problem for ϕ is linear and independent of t, with the non-linearity intruding only in the integration of eqn (1.28) with respect to time to obtain the pressure. For two-dimensional flows a stream function ψ exists and $\nabla^2 \psi = 0$, so that a complex potential $w = \phi + i\psi$ may be defined and a variety of simple solutions developed by complex variable methods.

The relevance of this inviscid irrotational model is restricted to regions in which viscosity is negligible. For many fluids viscosity is very small, but nevertheless cannot be neglected in a region adjacent to a solid boundary, where a boundary layer occurs across which the tangential velocity rapidly adjusts from zero to its predicted inviscid value. Vorticity is however generated in this boundary layer and is convected with the fluid particles, so that a wake is formed in which the vorticity is not zero. For streamlined bodies such a wake is thin and does not materially affect the overall flow. For a bluff body, however, the boundary layer will separate from its surface to form an extensive wake which will alter the irrotational flow regions. Thus the inviscid flow model, without any further modifications, is only relevant for flow past bodies which are streamlined to give thin wakes. It is also relevant for free-boundary flow, that is flow adjacent to a region of constant pressure. At such a boundary, whose position will be unknown and part of the problem, the relative normal velocity will be zero and the pressure will be prescribed from the continuity of normal stress. The tangential stress is necessarily zero from the neglect of viscosity, and no adjustment layer is needed at such a boundary.

If a free boundary $z = \eta(x, y, t)$ models the surface of a large extent of deep water moving under gravity (in the negative z

direction), then the boundary-value problem for ϕ and η is

$$\nabla^2\phi = 0, \quad -h < z < \eta;$$

$$z = -h, \quad \frac{\partial\phi}{\partial z} = 0;$$

$$z = \eta, \quad \frac{\partial\phi}{\partial z} = \frac{\partial\eta}{\partial t} + \frac{\partial\phi}{\partial x}\frac{\partial\eta}{\partial x} + \frac{\partial\phi}{\partial y}\frac{\partial\eta}{\partial y},$$ (1.29)

$$\frac{\partial\phi}{\partial t} + \tfrac{1}{2}(\nabla\phi)^2 + g\eta = G(t);$$

where for simplicity h is constant and initial conditions are given. This is a difficult problem because of the non-linear boundary conditions, but a solution does exist for a limited time. If, however, the surface disturbances are small so that the conditions can be linearized, then eqn (1.29) reduces to the very successful model for water waves in the form

$$\nabla^2\phi = 0, \quad -h < z < 0;$$

$$z = -h, \quad \frac{\partial\phi}{\partial z} = 0; \quad z = 0, \quad \frac{\partial^2\phi}{\partial t^2} + g\frac{\partial\phi}{\partial z} = 0;$$ (1.30)

where $z = 0$ is the mean surface height. For details of the relevance and use of this model see Ockendon and Tayler (1983).

In both the viscous and inviscid models the energy equation (1.3) has not interacted with the continuity and momentum equations, that is the problem may be solved for the flow without regard to the energy equation. For the inviscid model eqn (1.3) reduces to

$$\rho c\left(\frac{\partial T}{\partial t} + \boldsymbol{q} \cdot \nabla T\right) = k\nabla^2 T,$$ (1.31)

and if the Peclet number is large the conduction term may be neglected (away from boundaries), and the temperature is constant for each fluid particle. Near a boundary on which the temperature is prescribed there will be a (temperature) boundary layer across which the temperature adjusts from the given value to that convected by the fluid.

For a compressible fluid the energy equation is part of the flow problem and we consider an ideal gas, which is inviscid, non-heat-conducting, and has a gas law $p = \rho RT$ where R is a constant. The equations of continuity, momentum and energy are in this case

$$\frac{\partial \rho}{\partial t} + \nabla(\rho \boldsymbol{q}) = 0, \tag{1.32}$$

$$\rho \frac{\mathrm{d}\boldsymbol{q}}{\mathrm{d}t} = -\nabla p, \tag{1.33}$$

$$\frac{\mathrm{d}S}{\mathrm{d}t} = 0, \tag{1.34}$$

where the entropy S is proportional to p/ρ^{γ} and γ is the constant ratio of specific heats in the gas. This is the isentropic flow model, and for flows from uniform conditions the entropy S is a universal constant and the flow is homentropic. This implies that $p = p(\rho)$, and eliminating p the problem reduces to four scalar equations for four unknowns $\boldsymbol{q}$ and ρ. Thus we expect a well-posed problem, at least for a limited time given suitable spatial and initial boundary conditions. The model is, however, highly non-linear, and only a few closed-form solutions can be found (see Ockendon and Tayler (1983)).

By assuming that there are only small disturbances $\boldsymbol{q}$ about a state of rest with density ρ_0 and pressure p_0, the problem may be linearized in the form

$$\frac{\partial \rho}{\partial t} + \rho_0 \nabla \boldsymbol{q} = 0 = \rho_0 \frac{\partial \boldsymbol{q}}{\partial t} + p'(\rho_0) \nabla \rho.$$

The disturbance flow is irrotational (if initially uniform) so that $\boldsymbol{q} = \nabla \phi$, and eliminating ρ

$$\frac{\partial^2 \phi}{\partial t^2} = p_0'(\rho_0) \nabla^2 \phi. \tag{1.35}$$

This is the equation for acoustic (sound)waves, that is the propagation of small pressure disturbances in the gas, and the speed of sound is $[p'(\rho_0)]^{\frac{1}{2}}$.

It is interesting to note that the three simplest models described above, namely eqn (1.27) for an inviscid incompressible fluid, eqn (1.35) for small-disturbance compressible flows, and eqn (1.31) for the temperature distribution when the flow is negligible, have governing equations which are the canonical second-order scalar partial differential equations of elliptic, hyperbolic and parabolic type, called Laplace's equation, the wave equation, and the diffusion equation respectively. Solutions of these equations provide valuable information about the applications of the models. The applicatons will, however, provide results which give information about the expected qualitative properties of those equations. Thus there is an interaction between proven results relating to the equations in the model, and to the observations of the application of the model, to the mutual benefit of both the pure theory and the application. Moreover the equation is likely to appear as a model for a different application, and an analogy may be drawn between the two applications through the properties of the equation.

(c) *Solid Mechanics.* For a solid which undergoes a small deformation $u(x, t)$ a momentum balance similar to eqn (1.22) gives

$$\rho \frac{\partial^2 u_i}{\partial t^2} = \rho F_i + \frac{\partial \sigma_{ij}}{\partial x_j}. \tag{1.36}$$

The constitutive law for linear elasticity is

$$\sigma_{ij} = \lambda \frac{\partial u_k}{\partial x_k} \delta_{ij} + 2\mu \varepsilon_{ij}, \tag{1.37}$$

where ε_{ij} is the strain tensor $\frac{1}{2}[(\partial u_i/\partial x_j) + (\partial u_j/\partial x_i)]$ and λ and μ are the elastic constants related to Young's Modulus and Poisson's ratio. Eliminating σ_{ij}, the equations of linear elasticity for the deformation u in vector form are

$$\rho \frac{\partial^2 u}{\partial t^2} = \rho F + (\lambda + 2\mu)\text{grad div } u - \mu \text{ curl curl } u. \tag{1.38}$$

It is interesting to compare this with the Navier–Stokes equations (1.1) and (1.2). Equation (1.38) is second-order in the time derivative, but is linear and a single-vector equation for the unknown vector u. Thus we would expect a well-posed problem to have u prescribed on a spatial boundary with u and $\partial u/\partial t$ given initially, and also expect that it would be easier to find solutions for this model of linear elasticity than for viscous flow. Both these expectations are confirmed in practice, except that for many of the most interesting elasticity applications it is the stress distribution which is more important than the displacement; with viscous flow, however, the velocity field is usually more useful than the stress distribution. This in general makes problems complicated, since there are six independent components of the symmetric stress tensor, and for simplicity we only consider the case of plane strain.

Plane strain is an equilibrium (time-independent) situation in which $u_3 = 0$ and u_1 and u_2 are functions of x_1 and x_2 only. From the constitutive equation (1.37)

$$\sigma_{31} = \sigma_{32} = 0, \quad \sigma_{33} = \frac{\lambda}{2(\lambda + \mu)}(\sigma_{11} + \sigma_{22}),$$

so that a stress normal to the plane of interest is necessary to retain the two-dimensional character of the displacement. From the remaining three constitutive relations, u_1 and u_2 may be eliminated to give the compatibility relation between σ_{11}, σ_{22} and σ_{12} $(= \sigma_{21})$ as in Exercise 1.8. This, together with the two steady-state momentum equations (1.36) for the x_1 and x_2 directions, completes the model for the three unknown stress components. Complex variable methods may be used to construct solutions, and for details see England (1971).

An alternative approach to the reduction of the general equations of linear elasticity to simpler geometries is to reformulate the problem from first principles. A particularly successful use of this approach is the theory of small displacements of thin rods or beams. Consider first longitudinal displacements $u(x, t)$ with a local stress $\sigma(x, t)$ normal to the cross-section in a long straight rod of small uniform cross-section A. A momentum balance along the rod then gives

$$\rho A \frac{\partial^2 u}{\partial t^2} = \left(\frac{\partial \sigma}{\partial x} + q\right)A,$$

where q is any body-force per unit volume along the rod and ρ is the density. The constitutive law for linear elasticity, that is for small displacements, is then Hooke's law $\sigma = E(\partial u/\partial x)$ where E is Young's modulus. Thus we obtain the equation

$$\rho \frac{\partial^2 u}{\partial t^2} = E \frac{\partial^2 u}{\partial x^2} + q, \tag{1.39}$$

for the motion of longitudinal waves in thin rods.

This can only be derived from the general equation of elasticity (1.36) as an approximation in which displacements u_2 and u_3 in the rod cross-section are small, but their derivatives with respect to x_2 and x_3 are not negligible. Then with no normal stress in the cross-section, from the general constitutive equation (1.37)

$$\lambda \operatorname{div} \mathbf{u} + 2\mu \frac{\partial u_2}{\partial x_2} = 0 = \lambda \operatorname{div} \mathbf{u} + 2\mu \frac{\partial u_3}{\partial x_3}.$$

Solving for u_2 and u_3 with $u(=u_1)$ independent of x_2 and x_3,

$$u_2 = -\frac{\lambda}{2(\lambda + \mu)} x_2 \frac{\partial u}{\partial x}, \quad u_3 = -\frac{\lambda}{2(\lambda + \mu)} x_3 \frac{\partial u}{\partial x}.$$

Hence $\sigma(=\sigma_{11}) = [\mu(3\lambda + 2\mu)/(\lambda + \mu)] \partial u/\partial x$, and Young's modulus E is related to the elastic constants by $E = \mu(3\lambda + 2\mu)/(\lambda + \mu)$. Note that the model is not truly one-dimensional in that significant displacement gradients in the cross-section necessarily occur. If these displacements had been assumed to be identically zero the constitutive relation would reduce to $\sigma = (\lambda + 2\mu) \partial u/\partial x$, giving a different value for Young's modulus because of the normal stresses which now exist in the cross-section.

If the rod is given a small transverse displacement in a plane, there will be a neutral line in that plane on which there is no extension or compression. Take local coordinates x_1 and x_2 along and normal to this line, and let $v(x, t)$ be its displacement as in Fig. 1.1. Then neglecting the shear σ_{12} in comparison with the normal stress σ_{11}, we have $u_1 = -x_2(\partial u_2/\partial x_1)$. Also from (1.37), as before, $\sigma_{11} = E(\partial u_1/\partial x_1) = -Ex_2(\partial^2 u_2/\partial x_1^2)$, and for small displacements we may replace u_2 by v and x_1 by x. The normal stress creates a moment about $x_1 = x_2 = 0$ whose resultant $M = -E \int x_2^2 \, dA \, (\partial^2 v/\partial x^2) = -EI(\partial^2 v/\partial x^2)$, and I is the moment of inertia of the cross-section about its neutral point defined by $\int x_2 \, dA = 0$.

This bending moment creates a mean shear τ across the rod given by $A\tau = \partial M/\partial x$, if rotational inertia is neglected. Resolving normal to the rod then gives

$$\rho A \frac{\partial^2 v}{\partial t^2} = \left(q + \frac{\partial \tau}{\partial x} \right) A \tag{1.40}$$

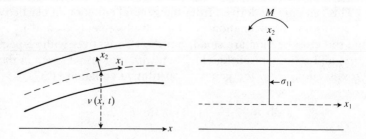

Fig. 1.1. Flexural vibration of a beam

where q is any body force per unit volume normal to the rod in the plane of the displacement. Thus

$$\rho\frac{\partial^2 v}{\partial t^2}+\frac{EI}{A}\frac{\partial^4 v}{\partial x^4}=q, \tag{1.41}$$

which is called the Bernoulli–Euler equation. The coefficient EI/A is the flexural rigidity, and is not necessarily small for a thin rod. The model has been modified to include the effects of rotational inertia by Timoshenko, and for details see Graff (1975) and Exercise 1.9. Appropriate boundary conditions for a well-posed problem will be, for example, that both the displacement and slope of the rod are given at both ends, in addition to initial conditions on displacement and velocity.

For a wire, both the variation in shear and transverse component of the stress on the cross-section may be important. If the wire is under tension T and there is a negligible displacement along the wire (and no body force in that direction), then from the momentum balance in the x direction σ is constant, equal to $-T/A$. In the transverse momentum-balance equation (1.40) there will be a contribution from the stress σ, since it is directed along the wire which for small displacements is inclined at an angle $\partial v/\partial x$ to the x axis. Thus eqn (1.41) becomes

$$A\rho\frac{\partial^2 v}{\partial t^2}+EI\frac{\partial^4 v}{\partial x^4}-T\frac{\partial^2 v}{\partial x^2}=q. \tag{1.42}$$

With negligible flexural rigidity this reduces to the well-known equation for transverse waves on a string.

Exercises

1.1. A viscous fluid is sheared between two parallel planes, distance h apart, whose constant relative velocity is U_0. Both plates are maintained at a temperature T_0. Show that an exact solution of eqns (1.1), (1.2) and (1.3) for a Newtonian viscous fluid exists in the form

$$\frac{u}{U_0} = \frac{y}{h}, \quad \frac{T - T_0}{T_0} = \frac{Br}{2} \frac{y}{h}\left(\frac{y}{h} - 1\right),$$

where y is measured normal to the planes and $Br = \mu U_0^2 / \kappa T_0$. If U_0 varies in time, under what conditions will this still be a good approximation to the flow? Say whether conditions can be such that T satisfies

$$\rho c \frac{\partial T}{\partial t} = \frac{\mu U_0^2}{h^2},$$

and give detailed reasons.

1.2. (a) Reconsider problem A (Furnace reaction analysis) when both $\varepsilon \ll 1$ and $\delta \ll 1$, and show that eqn (A6) becomes

$$\alpha_0(\tau) = \exp[-A e^{-1/\varepsilon}(\varepsilon/\delta)(e^{(\delta\tau/\varepsilon)} - 1)] + \text{smaller terms}.$$

Show further that T_1 at $x = 0$ will have a maximum value T_m if there is a positive root τ_m of

$$A e^{-1/\varepsilon} \varepsilon\, e^{\delta\tau/\varepsilon} = \delta.$$

If $\delta > A\varepsilon e^{-1/\varepsilon}$, find an approximate expression for τ_m.

(b) If the sample of Problem A is in a box with constant temperature at both ends, and the rate equation may be modelled by $\partial\alpha/\partial t = -A e^{bT}$, show that a steady-state temperature distribution satisfies

$$\theta'' + c^2 e^\theta = 0; \quad \theta(1) = 0 = \theta'(0),$$

where $c^2 = (AL^2 \rho\lambda b/k)e^{bT_0}$, and θ is a suitably non-dimensional temperature.

If $\theta(0) = \theta_0$ and $s_0^2 = e^{\theta_0}$, show that $s_0 = \cosh(cs_0/\sqrt{2})$. Hence deduce that there is no solution for c greater than some critical value. (This crudely models the phenomenon of self-ignition.)

1.3. (a) Show that the Laplace transform $\hat{f}$ of a function f satisfying Bessel's equation $xf'' + f' + xf = 0$ has the form $A(1+p^2)^{-\frac{1}{2}}$ where A is a constant. By using the inversion integral, and reducing it to an integral between the branch points $p = \pm i$, show that the solution of Bessel's equation J_0 which has a Laplace transform and equals 1 when $x = 0$ is

$$\frac{2}{\pi} \int_0^1 \frac{\cos xs \, ds}{(1-s^2)^{\frac{1}{2}}} .$$

(b) Obtain a second solution by considering a suitable contour enclosing only one branch point, and by expanding about the branch point show that for large x (Laplace's method) its asymptotic form is $e^{\pm i(x-\pi/4)}/x^{\frac{1}{2}}$, the sign depending on which branch point is chosen (Hankel function). Hence show that $J_0(x) \sim (2/x\pi)^{\frac{1}{2}}\cos(x - \pi/4)$ as $x \to \infty$.

From the differential equation show that a Hankel function has a logarithmic singularity as $x \to 0$.

1.4. (a) A finite Fourier sine transform may be defined by

$$F_n(k) = \int_0^\pi f(x)\sin(nkx) \, dx.$$

Using the Fourier series theorem show that the inversion formula is

$$f(x) = \frac{2}{\pi} \sum_1^\infty F_n(k)\sin(nkx).$$

What is the corresponding result for the finite cosine transform?

The general second-order linear equation

$$\frac{d}{dx}\left(p\frac{df}{dx}\right) + (\lambda r - q)f = 0$$

with homogeneous boundary conditions on $x = 0$ and 1 has eigenvalues λ_n $(n = 1, 2, \ldots)$ and eigenfunctions $f_n(x)$. Show that if $g(x) = \sum_{n=1}^\infty c_n f_n(x)$ then

$$c_n = \int_0^1 r(s)g(s)f_n(s) \, ds.$$

If we write $c_n = \bar{g}(n)$, then $\bar{g}(n)$ is the eigenfunction transform of $g(x)$ and the inversion formula is $g(x) = \sum \bar{g}(n) f_n(x)$. If

$$\frac{d}{dx}\left(p\frac{df}{dx}\right) + (\lambda r - q)f = g,$$

and λ is not an eigenvalue, show that

$$c_n = \frac{1}{\lambda - \lambda_n} \int_0^1 g(s) f_n(s) \, ds.$$

Hence derive the Green's function for this problem in the form

$$G(x, s) = \sum_{n=1}^{\infty} \frac{f_n(s) f_n(x)}{\lambda - \lambda_n}.$$

(b) If f satisfies

$$xf'' + f' + xf = xg(x),$$

show that an integral representation is

$$f = \int_0^{\infty} s(1 - s^2) J_0(sx) G(s) \, ds,$$

where G is the Hankel transform of g.

1.5. (a) By considering the integral

$$\int_a^b \left[G(x, s) L\left(\frac{d}{ds}\right) f(s) - f(s) M\left(\frac{\partial}{\partial s}\right) G \right] ds$$

show that an alternative definition of G for eqn (1.16) is, in the case $\lambda = \mu = 0$,

$$M\left(\frac{\partial}{\partial s}\right) G = \frac{\partial^2 G}{\partial s^2} - \frac{\partial}{\partial s}(c_1(s)G) + c_2(s)G = \delta(s - x)$$

with $G = 0$ at $s = a$ and b. (M is the adjoint operator to L, and clearly for a self-adjoint equation G is symmetric and $c_1 \equiv 0$.)

(b) Given that $f(x)$ is real-valued continuous function on the interval $-1 \leq x \leq 1$, determine a Green's function $G(x, s)$ such that the equation

$$y'' = f(x) \quad (-1 \leq x \leq 1),$$

together with the boundary conditions $y(-1) = y(1) = 0$, has the solution

$$y(x) = \int_{-1}^{1} G(x, s)f(s)\, ds.$$

If the boundary conditions are now changed to

$$y(1) = y(-1) \quad \text{and} \quad y'(1) = y'(-1),$$

show that the boundary-value problem has a solution only if f has mean value zero over $(-1, 1)$, and in this case show that there is a solution in the form

$$y(x) = \int_{-1}^{1} K(x, s)f(s)\, ds,$$

where

$$K(x, s) = \tfrac{1}{2}|x - s| - \tfrac{1}{4}(x - s)^2.$$

State whether this is the only solution, justifying your answer.

1.6. (a) If the kernel $K(x, s)$ of the Fredholm integral equation (1.19) is separable, that is it may be written in the form

$$\sum_{r=1}^{n} a_r(x)b_r(s),$$

show that the problem reduces to solving

$$(\lambda I - D)\mathbf{c} = \mathbf{k},$$

where the components of $\mathbf{c}$ and $\mathbf{k}$ are

$$\int_{0}^{1} f(s)b_i(s)\, ds$$

and

$$\int_{0}^{1} k(s)b_i(s)\, ds$$

respectively, and the components of D are

$$\int_{0}^{1} b_i(s)a_j(s)\, ds.$$

Discuss the Fredholm alternative in this situation; and if D is symmetric and λ is an eigenvalue, obtain a necessary condition on k for there to be a solution.

(b) Show that the positive eigenvalue of the kernel $K(x, s) = \cosh(x - s)$ $(0 \leqslant x, s \leqslant 1)$ is $\frac{1}{4}(2 + e - e^{-1})$. Find a necessary condition on f for

$$u(x) = f(x) + \frac{4}{2 + e - e^{-1}} \int_0^1 \cosh(x - s)u(s)\,ds$$

to have a solution and find all the solutions when f satisfies the condition.

(c) For a symmetric kernel which is not necessarily separable show that the eigenvalues are real and that the eigenfunctions corresponding to distinct eigenvalues are orthogonal.

1.7. Evaluate asymptotically

$$I(x) = \int_0^\infty \frac{e^{-t}\,dt}{1 + xt},$$

when x is small. For x large and positive show that

$$I \sim \frac{1}{x}\,e^{1/x}\log x.$$

1.8. For plane strain ($u_3 = 0$, $\boldsymbol{u}$ is independent of x_3) show by direct elimination from eqn (1.37) that the stress components must satisfy

$$4(\lambda + \mu)\frac{\partial^2 \sigma_{12}}{\partial x_1\,\partial x_2} = (\lambda + 2\mu)\left(\frac{\partial^2 \sigma_{11}}{\partial x_2^2} + \frac{\partial^2 \sigma_{22}}{\partial x_1^2}\right) - \lambda\left(\frac{\partial^2 \sigma_{22}}{\partial x_2^2} + \frac{\partial^2 \sigma_{11}}{\partial x_1^2}\right).$$

In the steady case, with no body force, show that the equations of equilibrium (1.36), and this equation of compatibility, are satisfied by

$$\sigma_{11} = \frac{\partial^2 A}{\partial x_2^2}, \quad \sigma_{22} = \frac{\partial^2 A}{\partial x_1^2}, \quad \sigma_{12} = -\frac{\partial^2 A}{\partial x_1\,\partial x_2}$$

where A is a scalar function such that $\nabla^4 A = 0$ (Airy stress function).

Show how to modify this definition of A to include a constant body force, and give conditions sufficient to make A unique.

1.9. Show that if rotational inertia is included in the model for the bending of a beam then

$$\tau = \frac{I}{A} \left(\rho \frac{\partial^2 \psi}{\partial t^2} - E \frac{\partial^2 \psi}{\partial x^2} \right),$$

where ψ measures the rotation of each small element. Using the Timoshenko argument that the slope $\partial v / \partial x$ of the line of centroids of the beam is given by $\psi + \gamma \tau$, show that v satisfies an equation of the form (in the case of no body force q)

$$\frac{A}{I} \frac{\partial^2 v}{\partial t^2} + \frac{E}{\rho} \frac{\partial^4 v}{\partial x^4} - (1 + \gamma E) \frac{\partial^4 v}{\partial x^2 \partial t^2} + \rho \gamma \frac{\partial^4 v}{\partial t^4} = 0.$$

Hence show that on a short time and length scale it is possible to propagate waves with speeds $(E/\rho)^{\frac{1}{2}}$ and $(1/\gamma \rho)^{\frac{1}{2}}$.

2 Wave motion

2.1. Transverse waves on a stretched string

The simplest example of wave motion is that of small transverse motions of a stretched string. If the string is uniform and stretched along the x axis, then the transverse displacement $y(x, t)$ satisfies

$$\rho \frac{\partial^2 y}{\partial t^2} = \frac{\partial}{\partial x} \left(T \frac{\partial y}{\partial x} \right). \tag{2.1}$$

This equation is the result of resolving the forces and mass-acceleration on a small element of the string in the y direction. Also ρ is the mass per unit length of the string, assumed constant, and T is the tension, which will be constant from resolving in the x direction. We define $c^2 = T/\rho$ and obtain the second-order wave equation in one space variable

$$y_{tt} = c^2 y_{xx}, \tag{2.2}$$

where suffices are used to denote differentiation. It is easily verified that the general solution is

$$y = f(x - ct) + g(x + ct),$$

where f and g are arbitrary functions which are twice differentiable. This represents a superposition of two constant profiles, called travelling waves, moving with speeds $\pm c$. Each of these profiles satisfies a first-order wave equation

$$y_t \pm c y_x = 0, \tag{2.3}$$

and c is called the *wave speed*.

With given initial data $y = F(x)$, $y_t = cG'(x)$, $-\infty < x < \infty$, at $t = 0$, the solution has the form

$$2y = F(x - ct) - G(x - ct) + F(x + ct) + G(x + ct), \tag{2.4}$$

and can easily be shown to be unique.

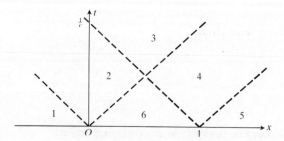

Fig. 2.1. Characteristic diagram for infinite string

If F and G' are both identically zero except in $0 < x < 1$, then eqn (2.4) may be illustrated by using an (x, t) diagram. Thus the functional form for y will differ in each of the six regions of the (x, t) plane in Fig. 2.1, defined by the lines $x \pm ct = $ constant through $(0, 0)$ and $(1, 0)$. These lines are the *characteristics* of the wave equation (2.1), and any discontinuity in the initial data at $x = 0$ and $x = 1$ will be *propagated* along them. One way of defining characteristics for second-order equations is that they are lines across which first derivatives in ϕ may have discontinuities, but it is only for *linear* second-order equations that this will imply that discontinuities in ϕ are propagated along characteristics. It is the existence of two characteristic directions at each point for a second-order equation which ensures that it is a hyperbolic problem, and that wave motion is possible.

In the above example the solution is zero in regions 1 and 5 because the initial data are zero for $x < 0$ and $x > 1$. This has the physical interpretation that for values of x greater than $(ct + 1)$ or less than $-ct$ the signal has not arrived, and the disturbance fronts or *wave fronts* are moving with speed $\pm c$ from $x = 1$ and 0 respectively. In deriving the solution in region 6, note that it is uniquely determined by the values of F and G in $0 < x < 1$; also note that in region 3, $G(x + ct) - G(x - ct) = G(0) - G(1)$ and is not equal to zero in general. We can now interpret an *infinite* string, for which eqn (2.4) is a solution, as a string whose end-points are far enough away from the initial disturbance that the wave fronts have not reached them in times of interest.

For a *semi-infinite* string one end is at $x = 0$ and, in the simplest case, is fixed so that $y(0, t) = 0$ for $t \geq 0$. The solution can be obtained from the infinite string problem which has initial data

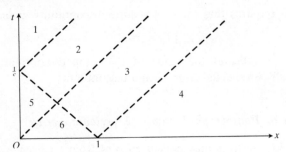

Fig. 2.2. Characteristic diagram for semi-infinite string

which are odd in x, so that the solution is odd in x and hence $y = 0$ on $x = 0$. Thus the characteristic diagram in Fig. 2.2 shows that there are still six regions to consider, and the characteristic of negative slope through $(1, 0)$ is *reflected* into a positive characteristic on $x = 0$. In regions 4 and 1, y will be identically zero, and regions 6 and 3 will not be affected by the boundary condition on $x = 0$. If y is prescribed on $x = 0$, say $y = K(t)$, where $K(0) = F(0)$, then to the solution with $y = 0$ on $x = 0$ we add $K(t - x/c)$ in regions 1 and 5 and satisfy all the boundary conditions.

Similarly if y_x was described on $x = 0$, the infinite string problem with even initial data would be used. However, for a mixed-type problem with $ay + by_x$ prescribed, this method of superposition fails and it is necessary to use other techniques. A unique solution is obtained to the half-space problem by these methods, and indicates that the boundary data are well posed if one is given y and y_t on $t = 0$, $x > 0$ and one condition on y and y_x on $x = 0$, $t > 0$. Since the wave equation (2.2) does not distinguish between variables x and ct, it is also clearly possible to have well-posed boundary data given y and y_x on $x = 0$, $t > 0$ and one condition on y and y_t on $t = 0$, $x > 0$. However, this would imply that information was propagated backwards in time along a characteristic $x + ct = $ constant. This is unacceptable for a *causal* model of a physical situation, and two boundary conditions may only be specified on $t = 0$, which is called a *spacelike* boundary.

With these simple ideas of superposition of solutions in different regions bounded by characteristics, called *characteristic domains*, we can consider a problem which concerns a moving force applied to a string. For a stationary force R, resolving in the y

direction requires that there is a slope discontinuity given by

$$[Ty_x]_-^+ = R,$$

where $[\ ]_-^+$ indicates the value of the jump discontinuity, and waves will be reflected from such a discontinuity.

Problem B. Pantograph dynamics I, periodic motion

Electric locomotives may collect current from an overhead wire through a mechanical device called a pantograph. This device must be designed in such a way that it stays in contact with the wire for all operating speeds of the locomotive, and in particular when it passes under one of the wire supports. The overhead-wire system may have a complicated geometry but the simplest model, which may be thought of as a single wire hanging under tension from regularly spaced stiff springs distant L apart, is likely to describe the worst performance of the pantograph. This too has a complicated structure but can be crudely modelled as a stiff spring and a dashpot, that is a damping mechanism linear in the velocity. The simplest situation is when the locomotive is moving at constant speed U so that the reaction $R(t)$ between the pantograph and the overhead wire is applied at $x = Ut$. Neglecting any horizontal motion and denoting the vertical displacement of the wire by $\bar{y}(x, t)$, then resolving in the absence of any supports

$$\rho(\bar{y}_{tt} + g) - T\bar{y}_{xx} = R(t)\,\delta(x - Ut),$$

where $\delta(x)$ is the Dirac delta function and we have neglected any flexural rigidity in the wire. This can be written in a more convenient form by subtracting out the static sag due to gravity, which is the shallow catenary $\bar{y} = \rho gx(x - L)/2T$, and making the variables non-dimensional. Thus if $y(x, t)$ is the vertical displacement relative to the static configuration made non-dimensional with maximum static sag, then

$$\frac{1}{c^2}\,y_{tt} - y_{xx} = \dot{P}\,\delta(x - t), \qquad (B1)$$

where c is the non-dimensional wave speed and $P = dP/dt = 8R/\rho g$, with $P(0) = 0$ for convenience.

This applied force $\dot{P}(t)$ will give rise to a moving discontinuity in the slope of the wire, and to calculate its magnitude eqn (B1) is written in coordinates $\xi = x - t$, $\tau = t$. The equation becomes

$$\frac{1}{c^2} y_{\tau\tau} - \frac{2}{c^2} y_{\tau\xi} + \left(\frac{1}{c^2} - 1\right) y_{\xi\xi} = \dot{P}(\tau)\, \delta(\xi). \tag{B2}$$

Integrating eqn (B2) with respect to ξ over a small interval containing $\xi = 0$ and assuming continuity of y, y_τ and $y_{\tau\tau}$, we obtain

$$\left(\frac{1}{c^2} - 1\right) [y_\xi]_{\xi=0-}^{\xi=0+} = \dot{P}(\tau).$$

In the original coordinates this condition is

$$[y_x]_-^+ = \frac{c^2}{1 - c^2} \dot{P}(t), \tag{B3}$$

and gives the magnitude of the slope discontinuity provided $c \neq 1$. This result may be obtained directly from a physical argument, resolving forces normal to the wire with the applied force $\dot{P}$ brought to rest, but is left to the reader as an exercise. The case $c \leq 1$ will be considered later; for the train, practical considerations imply that $c > 1$.

To model the pantograph a first step is to ignore its mass and consider that it is exerting a constant force together with a damping action, in which case

$$\dot{P}(t) = a - b\dot{Y}, \tag{B4}$$

where Y is the vertical displacement of the pantograph, determined when contact is maintained by

$$Y(t) = \bar{y}(t, t) = y(t, t) + 4t(t - 1). \tag{B5}$$

For more detailed pantograph dynamics reference may be made to Ockendon and Tayler (1983). The constants a and b represent the force exerted by the stiff spring and the dashpot strength respectively.

Consider first a motion in which the wire supports are far apart and the train impulsively attaches to the wire at time $t = 0$ at $x = x_0$, with the wire at rest. Outgoing waves will be propagated from $x = x_0$ and the characteristic domains of interest are shown in Fig. B1, with $y \equiv 0$ in regions 1 and 4. For regions 2 and 3

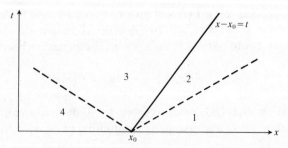

Fig. B1. Characteristic diagram for moving source

$y_2 = f_2[t - (x - x_0)/c]$ with $f_2(0) = 0$, and $y_3 = g_3[t + (x - x_0)/c]$ with $g_3(0) = 0$. On $x - x_0 = t$, the train path, the continuity of y and eqn (B3) give

$$y(t, t) = f_2\left[\left(1 - \frac{1}{c}\right)t\right] = g_3\left[\left(1 + \frac{1}{c}\right)t\right],$$

and

$$\frac{c^2 P(t)}{c^2 - 1} = \frac{1}{c - 1} f_2\left[\left(1 - \frac{1}{c}\right)t\right] + \frac{1}{c + 1} g_3\left[\left(1 + \frac{1}{c}\right)t\right],$$

where the second condition has been integrated once. From the contact condition (eqn B5) and the pantograph relation (eqn B4)

$$\dot{P}(t) = a - b\left[8t - 4 + \frac{c}{2}\dot{P}(t)\right].$$

The reaction is therefore a linear function of time and the pantograph displacement will be a quadratic function of time, for times such that neither wave front has been reflected by the overhead supports. This result is not particularly interesting but does demonstrate the qualitative effects of different pantograph dynamics. Thus if the pantograph is a spring providing a force $a_0 - a_1 Y$ and there is a no damping, then

$$\dot{P}(t) = a_0 - a_1\left[\frac{c}{2} P(t) + 4t^2 - 4t\right],$$

and the pantograph displacement is exponential in time.

A more interesting problem is to investigate the solution after a long time, rather than the initial motion, and as a first approximation assume that the supports are rigid so that the motion in

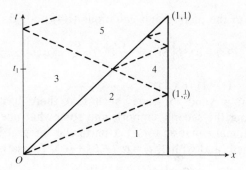

Fig. B2. Characteristic diagram for moving source with reflection

each span is independent of all other spans. That is we seek a periodic solution and need only consider $0 < x < 1$, with boundary conditions $y(0, t) = 0 = y(1, t)$ and $y(x, 0) = 0$ for $t < 0$, with the train arriving at $x = 0$ at time $t = 0$. The various characteristic domains are shown in Fig. B2 for the case $1 < c < 3$. Larger values of c would require consideration of further reflections using the same ideas. For this case there are five domains in which the wave equation (B1) has a solution in different functional forms, but we should also expect some distinction in our solution for P in region 4 due to the characteristic reflecting from the train path, as indicated by the dashed line. We have $y_1 = 0$ and $y_2 = f_2(t - x/c)$ where $f_2(0) = 0$; $y_3 = f_3(t - x/c) - f_3(t + x/c)$ satisfies $y = 0$ on $x = 0$; and $f_3(0) = 0$ from conditions at $x = t = 0$.

Now applying the continuity of y and eqn (B2) on $x = t$ for $0 < t \leqslant 2/(1+c) = t_1$, we have

$$f_2\left(\frac{c-1}{c}t\right) = f_3\left(\frac{c-1}{c}t\right) - f_3\left(\frac{c+1}{c}t\right),$$

$$\frac{c^2\dot{P}(t)}{1-c^2} = \frac{1}{c}f_3'\left(\frac{c-1}{c}t\right) + \frac{1}{c}f_3'\left(\frac{c+1}{c}t\right) - \frac{1}{c}f_2'\left(\frac{c-1}{c}t\right).$$

This second condition may be integrated with respect to t, and then eliminating f_2,

$$cP(t) = -2f_3\left(\frac{c+1}{c}t\right).$$

Substituting in the pantograph equations (B4) and (B5),

$$\dot{P}(t) = a - b\left[8t - 4 + \frac{c}{2}\dot{P}(t) - \frac{\mu c}{2}\dot{P}(\mu t)\right], \tag{B6}$$

where $0 < \mu = (c-1)/(c+1) < 1$.

This solution is valid for times such that there has been no reflection from the second support, and somewhat unexpectedly gives a functional equation for $\dot{P}$. A particular integral $\alpha_0 t + \beta_0$ is easily obtained, and with $\dot{P}(t) = \alpha_0 t + \beta_0 + f(t)$, $f(t)$ satisfies

$$(2 + bc)f(t) = b\mu c f(\mu t). \tag{B7}$$

This could be transformed into a more conventional functional equation in delay form by writing $t = e^z$ but a direct approach indicates that the only solution bounded at $t = 0$ is $f = 0$. For if a non-zero solution did exist such that $f = \gamma$ at $t = t_0$, then at $t_n = \mu^n t_0 < t_0$, $f = [(2+bc)/b\mu c]^n \gamma$. As $n \to \infty$, $t_n \to 0$ and f becomes unbounded.

We conclude therefore that in this time interval the reaction is a linear function of t, as in the case of initial motion but with different coefficients. For $t > 2(1+c)$,

$$y_4 = f_2\left[t - \frac{x}{c}\right] - f_2\left[t + \frac{x-2}{c}\right]$$

and y satisfies the condition $y_4(1, t) = 0$ and $y_4 = y_2$ on $x + ct = 2$. Also

$$y_5 = f_3\left[t - \frac{x}{2}\right] + g_5\left[t + \frac{x-2}{c}\right]$$

where $g_5(0) = -f_3(2/c)$, since $y_5 = y_3$ on $x + ct = 2$. A functional equation for $\dot{P}$ is again obtained but its solution will not be discussed further here. $\dot{P}$ will again be a linear function of t whose coefficients change in different time intervals corresponding to reflections of the characteristic $x = ct$ from the train path $x = t$ and boundary $x = 1$, shown as a dashed line in Fig. B2. These coefficients will be bounded in the limit as these time intervals decrease to zero as $x \to 1$ provided that $b < \frac{1}{2}[(c-1)/c]^2$, which is therefore an important design constraint (see Exercise 2.1).

If the pantograph had been modelled as a spring, then for

$0 < t < 2/(1+c)$ the equation for P becomes

$$\dot{P}(t) = a_0 - a_1 \left[4t^2 - 4t + \frac{c}{2} P(t) - \frac{c}{2} P(\mu t) \right].$$

This also has a particular solution for $P(t)$ which is quadratic in t, but now any homogeneous solution satisfies the functional differential equation

$$\dot{f}(t) = \frac{a_1 c}{2} [f(\mu t) - f(t)].$$

This is a special case of the functional differential equation

$$\frac{\mathrm{d}}{\mathrm{d}t} f(t) = k_1 f(\mu t) + k_2 f(t), \qquad \text{(B8)}$$

and a general discussion of such equations is given in Fox *et al.* (1971). A perturbation solution for μ approximately equal to unity will be discussed as an example of perturbation methods in Chapter 5. Thus in the case of the real problem for the train, a straightforward application of classical mechanics has led to a mathematical model which requires the solution of an unusual mathematical problem of considerable interest. Results may be obtained by numerical methods which are in reasonable agreement with available data and the effects of altering the train speed and pantograph characteristics are easily predicted.

2.2. Second-order hyperbolic equations and weak solutions

The significant property of the wave equation is the existence of two characteristic directions at each point of the (x, t) plane, where in the previous section characteristics were introduced as lines across which discontinuities in the second derivatives of ϕ could be propagated. For the general second-order linear partial differential equation with two independent variables x and y, namely

$$A\phi_{xx} + 2B\phi_{xy} + C\phi_{yy} + D\phi_x + E\phi_y + F\phi = G, \qquad (2.5)$$

where A, B, C, D, E, F, and G are functions of x and y only, characteristic directions will exist if the relations

$$A[\phi_{xx}] + 2B[\phi_{xy}] + C[\phi_{yy}] = 0$$
$$\dot{x}[\phi_{xx}] + \dot{y}[\phi_{xy}] = 0 = \dot{x}[\phi_{xy}] + \dot{y}[\phi_{yy}]$$

define non-trivial values for the jump discontinuities $[\phi_{xx}]$, $[\phi_{xy}]$, and $[\phi_{yy}]$. This will be so if the determinant of their coefficients vanishes, that is

$$A\dot{y}^2 - 2B\dot{x}\dot{y} + C\dot{x}^2 = 0. \tag{2.6}$$

Thus two characteristic directions exist if $B^2 > AC$ and eqn (2.5) is *hyperbolic*.

For the wave equation it is easily verified that the characteristics are $\xi = t - (x/c) = \text{constant}$ and $\eta = t + (x/c) = \text{constant}$. In general if $\xi(x, y) = \text{constant}$ and $\eta(x, y) = \text{constant}$ define the characteristic families obtained by integrating the two roots of eqn (2.6), then the transformation into ξ, η variables reduces eqn (2.5) to the canonical form

$$L(\phi) = \phi_{\xi\eta} + a\phi_\xi + b\phi_\eta + d\phi = e, \tag{2.7}$$

where a, b, d, and e are functions of ξ and η only. A well-posed problem results if Cauchy data, that is ϕ and grad ϕ, are given on an initial spacelike curve and we demonstrate this by introducing the Riemann function u. Note that for the wave equation a, b, d, and e are all zero and the general solution is $\phi = f(\xi) + g(\eta)$ with initial data in time given on $\xi + \eta = 0$.

Define a function $u(\xi, \eta; \bar{\xi}, \bar{\eta})$ such that

$$L^*(u) = u_{\xi\eta} - (au)_\xi - (bu)_\eta + du = 0 \tag{2.8}$$

in the domain $\xi < \bar{\xi}$, $\eta < \bar{\eta}$, with

$$u = \exp \int_{\bar{\xi}}^{\xi} b \, d\xi \quad \text{on} \quad \eta = \bar{\eta},$$

$$u = \exp \int_{\bar{\eta}}^{\eta} a \, d\eta \quad \text{on} \quad \xi = \bar{\xi}. \tag{2.9}$$

Then using Green's identity over the domain D between the

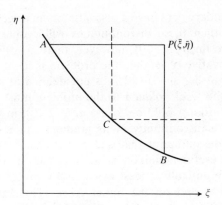

Fig. 2.3. Domains of influence and dependence

characteristics through P and the boundary curve (see Fig. 2.3),

$$\iint_D [uL(\phi) - \phi L^*(u)] \, d\xi \, d\eta = \int_{\partial D} (ua\phi + \lambda u\phi_\eta - \mu\phi u_\eta) \, d\eta$$

$$- \int_{\partial D} (ub\phi + \mu u\phi_\xi - \lambda\phi u_\xi) \, d\xi$$

where $\lambda + \mu = 1$. Rearranging this identity,

$$\phi(\bar{\xi}, \bar{\eta}) = \iint_D ue \, d\xi \, d\eta + \lambda u_B \phi_B + \mu u_A \phi_A$$

$$- \int_A^B (ua\phi + \lambda u\phi_\eta - \mu\phi u_\eta) \, d\eta + \int_A^B (ub\phi + \mu u\phi_\xi - \lambda\phi u_\xi) \, d\xi.$$

$$(2.10)$$

If u can be computed (and it can be proved that it exists and is unique), then the solution of eqn (2.7) is given by eqn (2.10) in terms of the Cauchy data on AB and an integral over D, the *domain of dependence of P*. This solution will be unique for all λ, μ such that $\lambda + \mu = 1$, and is stable to small changes in the boundary data or boundary shape; that is, the problem is well posed. Conversely the boundary data at C influence the solution only in the quadrants not containing the boundary curve, and if we associate positive directions with ξ and η increasing there is only one such quadrant, the *domain of influence* of C. The difficulty in this approach lies in finding u (see Exercise 2.3).

If the boundary data have a discontinuity in a second derivative of ϕ at C, then these discontinuities will propagate along the characteristic through C. If, however, there is a discontinuity in the first derivative of ϕ, then the problem is incompletely stated since the term $\phi_{\xi\eta}$ in eqn (2.7) is not defined at such a discontinuity and the *weak solution* to the problem must be defined. A possible definition would be to use eqn (2.10), in which case ϕ_P would have a discontinuity in its gradient whenever ϕ_A or ϕ_B have such a discontinuity, that is along the characteristics through C. A more useful definition is to introduce a test function ψ which is differentiable at least twice, and vanishes identically for large ξ and η and on the initial curve. Then an appropriate definition of the weak solution with zero initial data is a function ϕ satisfying

$$\iint \phi L^*(\psi)\,\mathrm{d}x\,\mathrm{d}t = \iint \psi e\,\mathrm{d}x\,\mathrm{d}t \qquad (2.11)$$

for all possible ψ, where the integrals are taken over all positive space (the 'positive' side of the initial curve). If ϕ does not have discontinuities in its first derivatives, then Green's identity gives

$$\iint \psi[L(\phi) - e]\,\mathrm{d}x\,\mathrm{d}t = 0,$$

so that $L(\phi) = e$ and we regain (2.7). For non-zero initial data an appropriate line integral along the initial curve has to be included in eqn (2.11).

If $\nabla\phi$ has discontinuities on a curve C, then in addition from Green's identity we have

$$\int_C (\lambda\psi[\phi_n]\,\mathrm{d}\eta - \mu\psi[\phi_\xi]\,\mathrm{d}\xi) = 0.$$

Thus discontinuities $[\phi_\xi]$ are possible on the characteristic $\xi = $ constant, and in $[\phi_n]$ on $\eta = $ constant. This definition of a weak solution leads therefore to a unique solution with discontinuities in $\nabla\phi$ on the characteristics through C. The magnitudes of the discontinuities satisfy the relations

$$\frac{\mathrm{d}}{\mathrm{d}\eta}[\phi_\xi] = [\phi_{\xi\eta}] = -a[\phi_\xi]; \quad \frac{\mathrm{d}}{\mathrm{d}\xi}[\phi_n] = -b[\phi_n] \qquad (2.12)$$

and hence are constant if $a = b = 0$. It is not at all clear how to

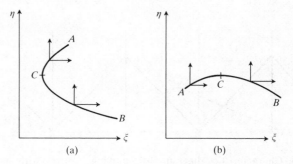

Fig. 2.4. Boundaries which are not space-like

define a weak solution which has discontinuities in ϕ, although a possible definition for the wave equation is to allow F and G of eqn (2.4) to be discontinuous functions. With this definition, discontinuities in ϕ will also propagate along the characteristics.

If the boundary curve has infinite (or zero) slope at a point C as in Fig. 2.4, then the arc AC is no longer spacelike and only one boundary condition can be prescribed on AC. For the wave equation ϕ_ξ is constant on lines $\xi = \text{constant}$ so that ϕ_ξ cannot be arbitrarily prescribed on AC in Fig. 2.4a, and likewise for ϕ_η in Fig. 2.4b.

For the wave equation the Riemann function u defined by eqns (2.8) and (2.9) with $a = b = d = 0$ is equal to unity, and eqn (2.10) gives an elegant method for determining the solutions $y(x, t)$ in the various regions of problem B. Thus from eqn (B1) with zero initial conditions

$$y(\bar{x}, \bar{t}) = \frac{c}{2} \iint_D \dot{P}(t)\, \delta(x - t)\, \mathrm{d}x\, \mathrm{d}t.$$

where we have not transformed to characteristic coordinates so that the Riemann function is $c/2$. The domain of integration D is bounded by the characteristics through $(\bar{x}, \bar{t})$ and the axis $t = 0$. To extend the finite domain of the problem we reflect the disturbance on $x = t$ in $0 < x < 1$ in the $x = 0$ and 1 boundaries as shown in Fig. 2.5, using alternate positive and negative values. Then integrating we obtain

$$y = \frac{c}{2} \int_\Gamma \dot{P}(t)\, \mathrm{d}t, \tag{2.13}$$

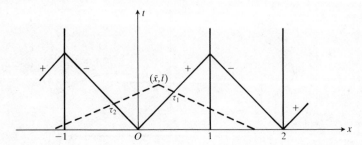

Fig. 2.5. Characteristic diagram for moving source with periodic reflections

where Γ is the section of the disturbance path and its reflection in the domain D. Thus to calculate y_3, for example, there are two contributions as shown in the figure, and

$$y_3 = \frac{c}{2}[P(\tau_1) - P(\tau_2)] = \frac{c}{2}\left[P\!\left(\frac{c\bar{t}+\bar{x}}{c+1}\right) - P\!\left(\frac{c\bar{t}-\bar{x}}{c+1}\right)\right],$$

confirming the result already obtained by elementary methods.

Problem C. Pantograph Dynamics II

We consider here two further aspects of the wave motion in the wire. First we note that the periodic solution obtained when $b < \frac{1}{2}[(c-1)/c]^2$ gives rise to a force $\dot{P}$, and hence a pantograph velocity $\dot{Y}$, which is discontinuous at each support. An improved model would be to recognize that the support is a stiff spring so that the boundary condition at each support is $[y]_-^+ = 0$, $\varepsilon[\partial Y/\partial x]_-^+ = y$, where $\varepsilon \ll 1$. Thus we define new variables $\bar{x}$ and $\bar{t}$ measured from the support and scaled with ε so that the problem is reduced to the train passing over a solitary support at $\bar{x} = 0 = \bar{t}$. The appropriate model is therefore, following eqn (B1),

$$\frac{1}{c^2}\, y_{\bar{t}\bar{t}} - y_{\bar{x}\bar{x}} = \dot{P}\delta(\bar{x} - \bar{t}) + y\delta(\bar{x}). \tag{C1}$$

The various solution domains are shown in Fig. C1, and

$$y_2 = \frac{c}{2}[P(\tau_1) + z(\tau)] = \frac{c}{2}\left[P\!\left(\frac{\bar{x}+c\bar{t}}{c+1}\right) + z\!\left(\frac{c\bar{t}-\bar{x}}{c}\right)\right],$$

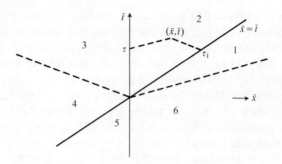

Fig. C1.

where

$$z = \int_0^{\bar{t}} y(0, \bar{t}) \, d\bar{t},$$

so that

$$\dot{z}(\bar{t}) = \frac{c}{2}\left[P\left(\frac{c\bar{t}}{c+1}\right) + z(\bar{t})\right], \quad \bar{t} > 0. \tag{C2}$$

For the pantograph defined by eqn (B4),

$$\dot{P}(\bar{t}) = -\frac{bc}{2}\frac{d}{d\bar{t}} y(\bar{t}, \bar{t}) + O(\varepsilon),$$

so that

$$P(\bar{t}) = -\frac{bc}{2}\left[P(\bar{t}) + z\left(\frac{c-1}{c}\bar{t}\right)\right].$$

Eliminating z,

$$(2 + bc)\left[P(\bar{t}) - \frac{2}{c-1}\dot{P}(\bar{t})\right] = bcP(\mu\bar{t}). \tag{C3}$$

Equation (C3) is valid in $\bar{t} > 0$ with $\mu < 1$, and has the form of eqn (B8); for $\bar{t} < 0$, the right hand side is replaced by $bcP(\bar{t}/\mu)$. It can be shown that the appropriate solution of these equations is $\dot{P} \equiv 0$ in $\bar{t} < 0$, with a 'pulse' in $\bar{t} > 0$, for which $\dot{P}(0) \neq 0$, and which matches together with piecewise linear solutions (obtained earlier) at the end of one period and the beginning of the next. For more general pantograph dynamics a system of equations of the form of eqn (B8) is obtained, leading to more complicated pulse shapes. The shape and maximum amplitude of the pulse is important, since for contact a necessary condition is $\dot{P} > 0$. These pulses will propagate and be reflected from further supports, the train path, and other pulses, but their amplitude diminishes at each reflection. Details of pulse shapes, and the resultant contact force profiles are given in Ockendon and Tayler (1983).

The second extension is to an accelerating train which can achieve speeds greater than the wire wave speed. As a simple example we take the train path to be $x = \frac{1}{2}t^2$ so that the wire wave speed is achieved at $t = c$, where a characteristic will be tangent to the train path at X as in Fig. C2, where no supports are included. The solution is obtained exactly as before and $y_1 = (c/2)P(\tau_2)$, where τ_2 is a root of $x - \frac{1}{2}\tau^2 = c(t - \tau)$ which is such that $\tau_2 < c$. Hence

$$y(t^2/2, t) = Y(t) = (c/2)P(t) \qquad 0 < t < c,$$
$$= (c/2)P(2c - t) \quad c < t < 2c.$$

This implies that $\dot{Y}$ is discontinuous at $t = c$, and for the pantograph modelled by eqn (B4), $\dot{P}$ is discontinuous at $t = c$ but is easily calculated. Now $y_2 = (c/2)P(\tau_1)$ and $y_3 = (c/2)[P(\tau_1) + P(\tau_2) - P(\tau_3)]$, so that the discontinuity in $\dot{P}$ at $t = c$ is propagated along the backward characteristic $x + ct = 3c^2/2$ through X, and an equal and opposite discontinuity is propagated along the forward characteristic $ct - x = c^2/2$. Thus the effect of

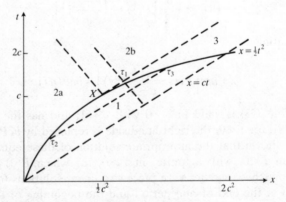

Fig. C2. Characteristic diagram for motion near a support

accelerating through the wire wave speed is to generate an impulse on the pantograph, which in this case propagates two slope discontinuities in the wire. However, in practice trains cannot easily achieve speeds comparable to the wire wave speed and we consider the related problem of an accelerating rocket hung from a taut spring for which some experimental data are available; see Rodeman, Longcope and Shampine (1976). The rocket is carried in a small cradle suspended from the wire, so that revolving vertically

$$\dot{P} = -a - b\ddot{Y}, \qquad (C4)$$

where a is due to gravity and b is the non-dimensional mass in our variables. If the rocket path is taken to be $x = \frac{1}{2}t^2$, then at $t = c$ there will be a discontinuity in $\dot{Y}$, that is a vertical impulse applied by the wire to the rocket. Equation (C4) then implies that there is a discontinuity in P at $t = c$, since its integrated form is $P = -at - b\dot{Y}$. A discontinuity in P implies that a discontinuity in y is propagated along the characteristics through X, that is the wire has a dramatic 'kink' and the assumption of small transverse slopes and displacements of the wire clearly fails. Thus the model will not be appropriate for the accelerating rocket near $t = c$.

Another illustration of the limitations of the model can be obtained by comparing the solutions for constant speed in the cases $c > 1$ and $c < 1$ as $c \rightarrow 1$ as in Exercise 2.2.

2.3. Linear hyperbolic systems in two independent variables

An alternative formulation of the wave equation is in the form of a pair of first-order equations

$$u_t - v_x = 0 = v_t - c^2 u_x \qquad (2.14)$$

where $u = \phi_x$ and $v = \phi_t$. A general linear first-order system may be defined by

$$L(\boldsymbol{u}) = A\boldsymbol{u}_t + B\boldsymbol{u}_x + C\boldsymbol{u} = \boldsymbol{d}, \qquad (2.15)$$

where A, B, and C are $n \times n$ matrices, $\boldsymbol{d}$ a column vector, all of whose components are functions of x and t only, and the solution

u has n components. Define an eigenvalue λ and a left eigenvector z^T (row vector) by

$$z^T(B - \lambda A) = 0,$$

so that λ is a root of $\det(B - \lambda A) = 0$. Then

$$z^T A(u_t + \lambda u_x) + z^T Cu = z^T d.$$

Hence on a curve defined by $dx/dt = \lambda$,

$$z^T A \frac{du}{dt} + z^T Cu = z^T d, \qquad (2.16)$$

where d/dt is differentiating with respect to time along the curve. If there are n real (distinct) roots for λ, the system is said to be hyperbolic and there will be n distinct relations like eqn (2.16), that is n first-order ordinary differential equations to determine u. These curves are the characteristics of the system, and a well-posed problem requires that u is given on some initial curve which intersects all the characteristics once. It is easily verified that the characteristics are the lines on which discontinuities $[u_t]$ and $[u_x]$ may occur, since

$$A[u_t] + B[u_x] = 0 = [u_t] + \lambda[u_x] \qquad (2.17)$$

implies $\det(B - \lambda A) = 0$.

If eqn (2.16) can be integrated (which for the linear problem normally requires A and B to have constant components), so that $w = p^T u + q = \text{constant}$ on the curve $dx/dt = \lambda$, then w is called a *Riemann invariant*. For the wave equation, written as a system according to eqn (2.14),

$$\det(B - \lambda A) = \begin{vmatrix} -\lambda & -1 \\ -c^2 & -\lambda \end{vmatrix},$$

so that $\lambda = \pm c$ gives the characteristics. The left eigenvectors are $(c, -1)$ and $(c, 1)$, with Riemann invariants $cu - v$ and $cu + v$.

A weak solution, for which u may be discontinuous, can be defined for the hyperbolic system of eqn (2.15) by introducing a test vector function ψ^T satisfying

$$L^*(\psi^T) = -(\psi^T A)_t - (\psi^T B)_x + \psi^T C$$

and vanishing far from the initial curve. Then the weak solution is a vector function u which for all admissible ψ^T satisfies, in the case of zero initial data,

$$\iint_D [L^*(\psi^T)u + \psi^T d]\, dx\, dt = 0. \tag{2.18}$$

This clearly reduces to eqn (2.15) if u is differentiable in D by using Green's identity on

$$\iint_D [\psi^T L(u) - L^*(\psi^T)u]\, dx\, dt,$$

and for non-zero initial data appropriate line integrals along the initial curve must be included in eqn (2.18). If u has a discontinuity across a curve C, then also

$$\int_C \psi^T (B[u]\, dt - A[u]\, dx) = 0.$$

Hence

$$(B - \lambda A)[u] = 0, \tag{2.19}$$

where $\lambda = dx/dt$. Thus the curves are characteristics, and possible discontinuities are given by eqn (2.19).

This result is not true for systems which are non-linear, and quasi-linear systems will be discussed in Section 2.7.

Problem D. A regenerative heat exchanger

This device is a rigid porous matrix through which two gases are alternately fed in opposite directions, and is commonly used in air-conditioning systems with the matrix rotating through the gas streams. Heat is removed from one gas by the matrix and used to heat the second colder gas. We are interested in the possibility of a stable periodic situation after a large number of cycles, and the consequent relationship between the period 2τ and the matrix temperature at the end of each half cycle. A simple model is required which highlights the heat transfer and the periodic nature of the process. Thus we avoid all discussion of the difficulties of gas flow in a porous medium, and assume that there will be an average motion for each gas which is constant, and either in the positive or negative x direction. Further we ignore all difficulties about the possible mixing of the two gases and assume that a pulse of hot gas is blown through the matrix with speed U in a half cycle. A more detailed discussion of the dynamics of gas flow through a porous matrix will be described later in Chapter 3.

Thus our model will be derived from a balance of heat, or of energy, for each gas and the matrix, and we shall ignore the effects of heat conduction in each gas, only considering the heat transfer to and from the matrix. This heat exchange may be crudely modelled by assuming that the local rate of heat exchange is $h(\theta_g - \theta_m)$, where θ_m and θ_g are the temperatures of the matrix and gas respectively and h is a heat-transfer coefficient which has to be determined empirically. Then a heat balance gives for the matrix

$$(1 - \mu)\rho_m c_m \frac{\partial \theta_m}{\partial t} = h(\theta_g - \theta_m),$$

and for either gas

$$\mu \rho_g c_g \left(\frac{\partial \theta_g}{\partial t} \pm U \frac{\partial \theta_g}{\partial x} \right) = h(\theta_m - \theta_g),$$

where μ is the voidage (gas space per unit volume) of the porous matrix, and alternate positive and negative signs occur, corresponding to the flow of hot and cold gas. To avoid ambiguity the suffix notation for derivatives has not been used here.

To these differential equations for θ_m and θ_g as functions of x and t must be added boundary conditions which are the given gas temperatures at the inlet ends of the matrix for each half cycle, and suitable initial conditions in the matrix. We use non-dimensional coordinates so that the matrix is $0 < x < 1$, the hot gas has inlet temperature 1, the cold gas $\theta_0 < 1$, and the initial matrix temperature is zero. We scale time so that the heat balance equation for the matrix reduces to

$$\frac{\partial \theta_m}{\partial t} = \theta_g - \theta_m, \tag{D1}$$

and for the gases

$$\pm \frac{\partial \theta_g}{\partial x} = \alpha(\theta_m - \theta_g) - \varepsilon \frac{\partial \theta_g}{\partial t}, \tag{D2}$$

where $\alpha = hL/\mu U \rho_g c_g$, $\varepsilon = hL/(1 - \mu)U\rho_m c_m$ and the gases may have different densities and specific heats.

The boundary conditions are $t = 0$, $\theta_m = 0$; with $x = 0$, $\theta_g = 1$, odd half cycles, $t > 0$; with $x = 1$, $\theta_g = \theta_0$, even half cycles, $t > 0$. Equations (D1) and (D2) form a linear first-order system in two dependent variables with characteristic directions $\lambda = dx/dt$ given by

$$\det\begin{pmatrix} -\lambda & 0 \\ 0 & \pm 1 - \lambda\varepsilon \end{pmatrix} = 0,$$

so that the characteristics are lines $x = \text{constant}$, and $t \mp \varepsilon x = \text{constant}$ (odd and even half cycles respectively). The boundary conditions are discontinuous at the end of each half-cycle but from eqn (2.19) θ_g is continuous across $x = \text{constant}$, and θ_m is continuous across $t \mp \varepsilon x = \text{constant}$ if we use the weak solution defined in eqn (2.18). Thus to complete the model we either give these continuity conditions or require that eqns (D1) and (D2) hold in integral form over any x, t domain.

A numerical approach will be necessary to solve the full equations (D1) and (D2) and it is important to be clear about the appropriate boundary data needed for a well-posed problem. We can do this formally by eliminating θ_m to obtain a second-order equation for θ_g in the form

$$\pm\frac{\partial^2 \theta_g}{\partial x\, \partial t} + \varepsilon\frac{\partial^2 \theta_g}{\partial t^2} \pm \frac{\partial \theta_g}{\partial x} + (\varepsilon + \alpha)\frac{\partial \theta_g}{\partial t} = 0. \tag{D3}$$

The Riemann solution of eqn (2.10) involves an unknown Riemann function u which would be difficult to evaluate in terms of simple functions, but it does demonstrate the dependence of the solution on the boundary conditions. Thus in Fig. D1 the gas flows occur in the parallelogram regions $1, 2, 3$ etc. bounded by characteristics through the points $(0, 0)$, $(1, \tau)$, $(0, 2\tau)$ etc. For regions $1, 3, 5$ etc. $\theta_g = 1$ on $x = 0$; and for regions $2, 4, 6$ etc. $\theta_g = \theta_0$ on $x = 1$. Initially $\theta_m = 0$ on the leading characteristic OA and the solution is determined in region 1 from the data on OA and OB. The region BCD is a result of the assumption that the two gases do not mix, that is the cold pulse is not switched on until the hot one is through the matrix and θ_m is constant in time in this region. Thus θ_m is prescribed on CD, θ_g on CE and the solution in region 2 is determined, to give θ_m on the leading characteristic of the next cycle. The problem is therefore well

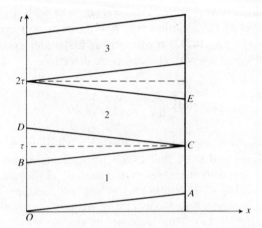

Fig. D1. Characteristic diagram for a regenerator

posed and a suitable finite difference scheme could be constructed. However, in practice the non-dimensional parameter ε
is small because the time for each gas to travel through the matrix
is short compared to the heat-transfer time scale. This simplifies
Fig. D1 since as $\varepsilon \to 0$ the parallelogram regions reduce to
rectangles, and the condition on $t = \tau, 2\tau, 3\tau$ etc. is that θ_m is
continuous. It also greatly simplifies any finite difference scheme,
which may however have to be used for a large number of cycles.

Further analytical progress can be made with the Riemann
solution if $\varepsilon = 0$. Making the transformation $\bar{\theta}_m = e^{t \pm \alpha x}\theta_m$, $\bar{\theta}_g = e^{t \pm \alpha x}\theta_g$, eqns (D1) and (D2) become

$$\frac{\partial \bar{\theta}_m}{\partial t} = \bar{\theta}_g, \quad \frac{\partial \bar{\theta}_g}{\partial x} = \pm \alpha \bar{\theta}_m, \tag{D4}$$

so that eqn (D4) reduces to the telegraphy equation

$$\frac{\partial^2 \bar{\theta}_g}{\partial x \, \partial t} = \pm \alpha \bar{\theta}_g, \tag{D5}$$

whose Riemann function u is known (Exercise 2.3).

Thus given that $\theta_m = T_{2n}(x)$ at $t = 2n\tau$, so that $\bar{\theta}_m = e^{2n\tau + \alpha x}T_{2n}(x)$ and $\partial \bar{\theta}_g / \partial x = \alpha e^{2n\tau + \alpha x}T_{2n}(x)$, the solution in the
half-cycle $2n\tau < t < (2n+1)\tau$ can be calculated from eqn (2.10),
using in addition that $\bar{\theta}_g = e^t$ on $x = 0$. For simplicity we do the

calculation in the case $\alpha = 0$ (which is unlikely to be true in practice) when the Riemann function is 1, and eqns (D4) integrate trivially as follows:

$$2n\tau < t < (2n+1)\tau;$$
$$\bar{\theta}_g = e^t, \quad \bar{\theta}_m = e^t - e^{2n\tau} + e^{2n\tau}T_{2n}(x).$$

Evaluating this on $t = (2n+1)\tau$,

$$T_{2n+1}(x) = 1 + (T_{2n}(x) - 1)e^{-\tau}.$$

Proceeding to the next half cycle, with $\bar{\theta}_g = \theta_0 e^t$ on $x = 1$,

$$(2n+1)\tau < t < (2n+2)\tau;$$
$$\bar{\theta}_g = \theta_0 e^t, \quad \bar{\theta}_m = \theta_0(e^t - e^{(2n+1)\tau}) + e^{(2n+1)\tau}T_{2n+1}(x).$$

Finally

$$T_{2n+2}(x) = \theta_0 + [T_{2n+1}(x) - \theta_0]e^{-\tau}.$$

A periodic motion is therefore possible with $T_{2n+2} = T_{2n}$ if

$$T_{2n} = \frac{\theta_0(e^\tau - 1) + 1 - e^{-\tau}}{e^\tau - e^{-\tau}} = T_\infty. \tag{D6}$$

This motion will in fact occur, since

$$T_{2n+2} - T_\infty = e^{-2\tau}(T_{2n} - T_\infty) = e^{-2n\tau}(-T_\infty)$$

tends to zero as $n \to \infty$, and the operating temperature of the matrix can be controlled by adjusting the period τ.

With $\alpha \neq 0$ an integral equation for possible periodic values of $T_{2n} = T_\infty$ will be obtained as in Exercise 2.3. Further information about the operation and construction of heat regenerators may be found by reference to Romie (1979) or Maclaine-Cross and Banks (1972).

2.4. Multiple reflections and the use of transforms

In the problems considered so far, characteristics have been reflected at boundaries $x = 0$ and $x = 1$ but there has been no need to consider more than a few such reflections. To obtain solutions for large times of initial-value problems on a finite x domain requires that there are many characteristic reflections, and the use of characteristic or Riemann methods is inappropriate. If an analytical solution of a given problem is to be possible, then usually the use of transforms is the most effective method. The most obvious approach for a hyperbolic problem with boundary data given on $x = 0$ and $x = 1$, and initial data on $t = 0$, is to use the Laplace transform

$$\hat{y} = \int_0^\infty e^{-pt} y(x, t) \, dt,$$

whose detailed properties may be found in many standard texts. We briefly illustrate its use by considering the non-homogeneous wave equation

$$\frac{1}{c^2} \frac{\partial^2 y}{\partial t^2} - \frac{\partial^2 y}{\partial x^2} = f$$

with boundary data $y = 0$ on $x = 0$ and 1; $y = F(x)$, $\partial y/\partial t = 0$ on $t = 0$. Then $\hat{y}$ satisfies

$$\frac{1}{c^2} [p^2 \hat{y} - pF(x)] = \frac{\partial^2 \hat{y}}{\partial x^2} + \hat{f},$$

with $\hat{y} = 0$ at $x = 0$ and 1.

To solve this ordinary differential equation for $\hat{y}$ we first obtain the Green's function $G(x, s)$ defined by

$$\frac{\partial^2 G}{\partial x^2} - \frac{p^2}{c^2} G = \delta(x - s); \quad G = 0 \quad \text{at} \quad x = 0, 1.$$

The solution then is

$$\hat{y} = - \int_0^1 G(x, s)[\hat{f}(s, p) + (p/c^2)F(s)] \, ds.$$

The inversion theorem for the Laplace transform, given by eqn (1.8), gives the solution for y in the form of a double integral which is often difficult to interpret. Thus an alternative approach is to use a finite transform in x, and the appropriate transform to be used will depend on the differential operator in the x variables and the form of the boundary conditions at $x = 0$ and 1. Thus for the example under discussion a finite Fourier sine transform is defined by

$$y_n(t) = \int_0^1 y(x, t) \sin n\pi x \, dx, \tag{2.20}$$

with inversion formula obtained from eqn (1.10) in the form

$$y(x, t) = 2 \sum_{n=1}^{\infty} y_n(t) \sin n\pi x. \qquad (2.21)$$

The boundary-value problem reduces to the ordinary differential equation

$$\frac{1}{c^2} \frac{d^2 y_n}{dt^2} + n^2 \pi^2 y_n = f_n(t), \qquad (2.22)$$

with $y_n = a_n = \int_0^1 F(x) \sin n\pi x \, dx$, $dy_n/dt = 0$ on $t = 0$. Hence

$$y_n = a_n \cos n\pi c t + \frac{n\pi}{c} \int_0^t f_n(\tau) \sin n\pi c(t - \tau) \, d\tau, \qquad (2.23)$$

and the solution is obtained using the inversion formula of eqn (2.21).

This solution in terms of an infinite sum of terms may also be obtained in a number of simple problems by the direct application of the technique of separation of variables, and superposition of solutions representing modes of oscillation. This elementary technique can be extended to other homogeneous boundary conditions on $x = 0$ and 1, and other differential operators with respect to x, and in each case there is a corresponding finite transform and inversion result. Many examples arise in classical problems of applied mechanics and we shall not discuss such techniques further here. Note, however, that if we had continued with the Laplace transform approach on a problem which could be solved by separation of the variables and superposition of modes, then the integrand of the complex inversion integral would have had an infinite series of poles, each of whose residues would have contributed one of the modes of the solution.

Problem E. Short circuit in a transformer

A transformer consists of concentric cylindrical primary and secondary windings attached to a rigid clamp structure on which the core is mounted, and the cross-section is shown in Fig. E1. Each winding consists of several hundred helical turns of thin copper strip separated by a thin strip of insulating pressboard. Hence in the cross-section of Fig. E1, each winding appears as a

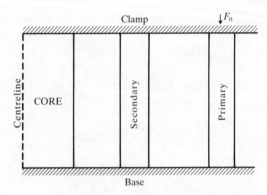

Fig. E1. Cross-section of a transformer

composite made up of alternate copper and pressboard elements. To give mechanical stability to this structure (somewhat analogous to a pack of cards) the windings are pre-stressed at their upper ends by a clamp. When a current flows in the transformer there is an electromagnetic coupling between the primary and secondary coils, giving rise to an electromagnetic field which exerts a mechanical force on the elements of each winding. The problem is to determine the mechanical response of a winding, say the secondary, under short-circuit conditions. The current in such conditions is known to have a time dependence similar to that in a circuit containing a resistance and an inductance to which an a.c. voltage is suddenly applied, that is a current proportional to $e^{-\alpha t} - \cos \omega t$ as shown in Fig. E2. The effects of such a short-circuit current can be powerful enough to buckle the secondary coil, which in an industrial context may be as much as 10 metres high with several hundred helical turns of copper. A

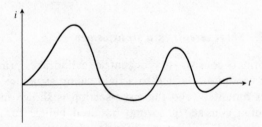

Fig. E2. Short circuit current

simple model is required which focuses on the mechanical effects, and an initial approach is to assume that the electromagnetic field in the winding under normal operating conditions is known, presumably by measurement.

The electromagnetic forces on the winding may then be assumed in the form of known measured functions of position and time multiplied by the square of the short-circuit current, at least for small displacements of the winding. These forces will act on each conducting element of the winding, which we would like to model as a composite bar made up of a large number of alternate conducting and elastic elements. That is, we would hope to obtain a continuous model for the bar, whose properties are continuous functions of space and time, by some limit process as the number of elements in a finite length increases.

Define a space variable x by the height of pressboard when no current flows, so that the corresponding height in the composite is $(1+\lambda)x$, where λ is the ratio of the thicknesses of the conductor and insulator and $0 < x < L$. When current flows the material element is displaced to a height $(1+\lambda)x + s(x, t)$, where s is a continuous function of x and t representing the expansion of the pressboard. Assume that there is an average stress $F(x, t)$ at each point in the pressboard, so that on applying Hooke's law

$$F - F_0 = -k \frac{\partial s}{\partial x}, \tag{E1}$$

where F_0 is the initial mechanical stress and k is a constant. Now F will be discontinuous across each conductor, so that resolving for an element δx which contains n conductors

$$\rho \frac{\partial^2 s}{\partial t^2} + \frac{\partial F}{\partial x} = \frac{\lambda}{n} \sum_{r=1}^{n} F_c(x, t)\, \delta(x - x_r), \tag{E2}$$

where F_c is the resultant force per unit length on each conductor of length $\lambda \delta x / n$, ρ is the linear density of the insulator, and gravity is neglected. Resolving for an individual conductor,

$$\rho_c \frac{\partial^2 s}{\partial t^2} = F_b - F_c, \tag{E3}$$

where F_b is the prescribed electromagnetic force per unit length and ρ_c is the linear density of the conductor. If F_c is not too rapidly varying, that is effectively constant over a number of conductors, then taking the limit of large n in eqn (E2) gives

$$\rho \frac{\partial^2 s}{\partial t^2} + \frac{\partial F}{\partial x} = \lambda F_c(x, t). \tag{E4}$$

Hence eliminating F_c and F we obtain

$$(\rho + \lambda \rho_c) \frac{\partial^2 s}{\partial t^2} - k \frac{\partial^2 s}{\partial x^2} = \lambda F_b, \tag{E5}$$

so that the wave speed c of longitudinal waves in the composite is given by $c^2 = k/(\rho + \lambda \rho_c)$.

If the base of the clamp is rigid and it is given a constant pre-stress, then suitable boundary conditions are $s = 0$ at $x = 0$, and $F = F_0$ (which implies $\partial s/\partial x = 0$ at $x = L$), together with zero initial data.

The crucial assumption in deriving this continuum model, by a process of *homogenization*, is that F_c varies slowly on length scales comparable with an individual conductor or insulator. However, F_c is not prescribed but depends on $\partial^2 s/\partial t^2$, although we can expect F_b to be sufficiently slowly varying. Thus slowly varying solutions like eqn (E5) will certainly be consistent, but other possibilities might conceivably occur (see Exercise 2.4).

We solve the model by taking a finite transform with respect to a function ϕ_n satisfying

$$\frac{1}{\phi_n} \frac{d^2 \phi_n}{dx^2} = \text{constant}$$

with $\phi_n(0) = 0 = (d\phi_n/dx)(L)$, that is $\phi_n = \sin[(n + \frac{1}{2})(\pi x/L)]$. Following eqns (2.20) and (2.21) we define

$$s_n = \int_0^L s \sin(n + \tfrac{1}{2}) \frac{\pi x}{L} \, dx,$$

where

$$s = \left(\frac{2}{L}\right) \sum_0^\infty s_n \sin(n + \tfrac{1}{2}) \frac{\pi x}{L} \, dx.$$

Then

$$\frac{d^2 s_n}{dx^2} + c^2 \frac{(n + \frac{1}{2})^2 \pi^2}{L^2} s_n = \frac{\lambda}{\rho + \lambda \rho_c} \int_0^L F_b \sin(n + \tfrac{1}{2}) \frac{\pi x}{L} \, dx = F_n(t). \tag{E6}$$

If the disturbance s is to be unexpectedly large, then the only possibility is for one of the $\bar{s}_n$ to become large. This is possible if eqn (E6) describes an oscillation near resonance, that is $F_n(t)$ contains a component of a harmonic of frequency $(n+\frac{1}{2})\pi c/L$. In practice the transformer is immersed in oil which acts as an efficient damping mechanism for high harmonics, but damage is likely to occur if the square of the current sketched in Fig. E2 has an appreciable component of a harmonic of frequency $(n+\frac{1}{2})\pi c/L$ for small values of n. The height L of the total winding can be changed by varying F_0, and hence resonant frequencies may be avoided if the short-circuit current is known.

2.5. Wave motion with two or more space variables

The wave equation naturally occurs with more than one space variable in many applications. For a stretched membrane whose undisplaced position is in the x, y plane, small displacements w in the z direction will be functions of x, y and t. Resolving in the x, y plane shows that the force per unit length, or tension T, on any small element must be constant. Resolving in the z direction for a small element area A, surface density σ, of the membrane,

$$A\sigma\frac{\partial^2 w}{\partial t^2} = \int T\frac{\partial w}{\partial n}\,\mathrm{d}s = T\int\nabla^2 w \,\mathrm{d}A \simeq TA\nabla^2 w.$$

Hence

$$\frac{\partial^2 w}{\partial t^2} = c^2\left(\frac{\partial^2 w}{\partial x^2} + \frac{\partial^2 w}{\partial y^2}\right), \tag{2.24}$$

the two (space) dimensional wave equation.

A general solution for eqn (2.24) does not exist as in the one dimensional case of eqn (2.1). Techniques of superposing general wave solutions in different domains are not available, and the use of characteristic domains is less valuable. Characteristics can still be defined as surfaces in (x, y, t) space across which discontinuities in the second derivatives of u are possible. For the wave equation (2.24) these characteristics are the cones $x^2 + y^2 - c^2 t^2 =$ constant, and for simple equations like this the methods of superposition of modes of vibration and of transforms may still be effective. Thus if we look for a solution of eqn (2.24) which is periodic in time with frequency $\omega(>0)$, then eqn (2.24) reduces to solving

$$\frac{\partial^2 w}{\partial x^2} + \frac{\partial^2 w}{\partial y^2} + \frac{\omega^2}{c^2} w = 0, \qquad (2.25)$$

together with boundary conditions at the perimeter of the membrane. Equation (2.25) is called Helmholz's equation and is elliptic, that is it has no real characteristic directions, and we delay general discussion of the solution of such problems to Chapter 3. Solutions do exist, as can be seen from the special case of a circular membrane of radius a, clamped at its perimeter so that $w = 0$ on $r = a$. Then by symmetry

$$\frac{d^2 w}{dr^2} + \frac{1}{r}\frac{dw}{dr} + \frac{\omega^2}{c^2} w = 0, \qquad (2.26)$$

subject to $w(0)$ bounded and $w(a) = 0$. This homogeneous problem has non-trivial solutions $w = \alpha J_0(\omega r/c)$, where α is an arbitrary constant and J_0 is a Bessel function, provided $\omega a/c$ is a zero of J_0, that is ω takes one of an infinite set of discrete values. We may reduce eqn (2.25) further by looking for solutions which are also periodic in x, so that

$$w = f(y)e^{i\omega t - ilx}, \qquad (2.27)$$

where we are using the convention that f may be complex and that taking the real part is implied. Then

$$\frac{d^2 f}{dy^2} + \left(\frac{\omega^2}{c^2} - l^2\right) f = 0,$$

and for $\omega^2 > c^2 l^2$ eqn (2.27) represents a harmonic plane wave $\alpha e^{i(\omega t - lx - my)}$ propagating in the (l, m) direction with speed c where $m^2 = \omega^2/c^2 - l^2$. For fixed l and m there are two roots for ω, corresponding to plane waves travelling in opposite directions. For an outgoing wave, by rotating the axes so that s is a variable measured outwards and normal to the wave front, the correct root for ω is chosen if

$$\frac{\partial w}{\partial t} + c\frac{\partial w}{\partial s} \to 0 \quad \text{as} \quad s \to \infty. \qquad (2.28)$$

This is called the *radiation condition* and clearly defines an outgoing wave for one-dimensional or plane waves. With circular symmetry eqn (2.24) reduces to

$$\frac{c^2}{r}\frac{\partial}{\partial r}\left(r\frac{\partial w}{\partial r}\right) = \frac{\partial^2 w}{\partial t^2}, \qquad (2.29)$$

and an outgoing wave would satisfy $\partial w/\partial t + c \, \partial w/\partial r \to 0$ as $r \to \infty$.

A simple example of a three-dimensional wave equation is as a model for acoustics, and may be obtained from the equations (1.32) and (1.33) for inviscid compressible flow by neglecting non-linear terms. Then q is irrotational and, with $c^2 = \gamma p_0/\rho_0$,

$$\phi_{tt} = c^2 \nabla^2 \phi. \qquad (2.30)$$

Solutions with spherical symmetry satisfy

$$c^2 \frac{\partial^2}{\partial r^2}(r\phi) = \frac{\partial^2}{\partial t^2}(r\phi), \qquad (2.31)$$

so that a general solution exists in the form

$$\phi = \frac{f(ct+r)+g(ct-r)}{r},$$

and an outgoing wave although not of constant profile, is clearly $\phi = g(ct-r)/r$.

A further example of three-dimensional waves arises from the equations of linear elasticity (1.38) derived in the form

$$\rho \frac{\partial^2 u}{\partial t^2} = (\lambda + 2\mu)\text{grad div } u - \mu \text{ curl curl } u,$$

where we now have a system of three equations with three space variables.

If we look for solutions in which curl $u = 0$ then

$$\rho \frac{\partial^2 u}{\partial t^2} = (\lambda + 2\mu)\nabla^2 u. \qquad (2.32)$$

These are called *longitudinal waves* since the displacement in a plane wave satisfying eqn (2.32) is in the direction of the wave (also called *P waves*). *Transverse waves* occur when div $u = 0$, so that

$$\rho \frac{\partial^2 u}{\partial t^2} = \mu \nabla^2 u \qquad (2.33)$$

and a two-dimensional wave equation is obtained if $u = [0, 0, w(x, y)]$, where the displacement is clearly normal to any propagation direction in the (x, y) plane (*SV waves*). For a transverse displacement in the (x, y) plane (*SH waves*), $u = (\psi_y, -\psi_x, 0)$, where ψ satisfies the two-dimensional wave equation, and the general transverse wave solution is a sum of these two displacements.

Problem F. Submarine detection

An important problem in submarine design concerns the control of the emission of sound waves into the ocean from the submarine hull. This is an example of a class of problems in which an oscillating membrane or elastic solid is immersed in a compressible fluid. Usually the fluid density is small enough for the fluid loading to act as a small perturbation to the basic membrane or plate oscillation which would occur in vacuo. These oscillations will be driven by the vibrations of the frame or ribs of the hull arising from the motion of the submarine's engines and shaft, and may be modelled by a periodic forcing term, frequency Ω.

For simplicity consider first a one-dimensional model in the z direction, in which the hull section is modelled as a spring-loaded piston, driven by a forcing term $F(t)$ but damped by a compressible fluid on one side of the piston. For small oscillations $w(t)$ about equilibrium,

$$\frac{d^2 w}{dt^2} + k^2 w = -\frac{p}{m} + F,$$

where $k^2 = \lambda/m$, λ being the modulus of the spring and m the mass per unit area of the piston. The fluid pressure on the piston is p, and if ϕ is the velocity potential of the fluid then

$$p = -\rho_0 \left(\frac{\partial \phi}{\partial t}\right)_{z=w}$$

and

$$\frac{dw}{dt} = \left(\frac{\partial \phi}{\partial z}\right)_{z=w}$$

For small motions of the piston these boundary conditions may be evaluated on $z = 0$, and reduce to

$$\frac{\partial^3 \phi}{\partial z\, \partial t^2} + k^2 \frac{\partial \phi}{\partial z} = \frac{\rho_0}{m} \frac{\partial^2 \phi}{\partial t^2} + \dot{F}. \tag{F1}$$

For an unbounded fluid in the positive z direction, ϕ satisfies eqn (2.30) and we require an outgoing wave satisfying eqn (2.28), so that $\phi = f(t - z/c)$. Substituting in eqn (F1), and integrating once, we obtain

$$f'' + \frac{c\rho_0}{m} f' + k^2 f = -F. \tag{F2}$$

With no forcing term F, there are periodic solutions of the form $f = e^{i\omega t}$ if

$$\omega^2 = k^2 + \frac{ic\rho_0}{m} \omega. \tag{F3}$$

This has no real roots for ω (k real), and for small fluid loading $\omega \simeq k(1 + i\varepsilon)$ where $\varepsilon = c\rho_0/2km \ll 1$. Thus the wave motion is damped (in time) by the presence of the fluid.

If, however, the piston is forced by a term $F_0 e^{i\Omega t}$, with frequency $\Omega = k[1 + O(\varepsilon)]$, then the appropriate solution is $f = A e^{i\Omega t}$ where A is complex and $O(1/\varepsilon)$. Thus a large-amplitude wave will be propagated into the fluid, and is likely to be detected.

To improve upon this model we consider a two-dimensional situation in which the hull section is modelled as a large membrane, surface density σ, which in equilibrium lies in the x, y plane under a uniform tension T, so that its displacement w satisfies eqn (2.24), supplemented by a fluid-loading and a forcing term. If we look for one-dimensional motions $w(x, t)$ in the membrane, we have

$$\frac{\partial^2 w}{\partial t^2} - c_1^2 \frac{\partial^2 w}{\partial x^2} = -\frac{p}{\sigma} + F,$$

where $c_1^2 = T/\sigma$. The velocity potential ϕ satisfies

$$c_2^2 \left(\frac{\partial^2 \phi}{\partial x^2} + \frac{\partial^2 \phi}{\partial z^2} \right) = \frac{\partial^2 \phi}{\partial t^2} \quad (z > 0),$$

where $c_2^2 = \gamma p_0/\rho_0$.

As in the one-dimensional case, $p = -\rho_0(\partial\phi/\partial t)$ and $\partial w/\partial t = \partial\phi/\partial z$ on $z = 0$, and we require only outgoing waves for large z. A general solution for ϕ is not possible in this two (space) dimensional situation, and we look for periodic solutions with frequency ω.

If $\phi = \tilde{\phi}e^{i\omega t}$ and $F = \tilde{F}e^{i\omega t}$, $(\omega > 0)$, the boundary-value problem for $\tilde{\phi}(x, z)$ is

$$\frac{\partial^2\tilde{\phi}}{\partial x^2} + \frac{\partial^2\tilde{\phi}}{\partial z^2} + \frac{\omega^2}{c_2^2}\,\tilde{\phi} = 0, \quad z > 0; \tag{F4}$$

$$\omega^2\frac{\partial\tilde{\phi}}{\partial z} + c_1^2\frac{\partial^3\tilde{\phi}}{\partial z\,\partial x^2} = \frac{\rho_0\omega^2}{\sigma}\,\tilde{\phi} - i\omega\tilde{F}, \quad z = 0; \tag{F5}$$

$$c_2\frac{\partial\tilde{\phi}}{\partial z} + i\omega\tilde{\phi} \to 0 \quad \text{as} \quad z \to \infty. \tag{F6}$$

This is an elliptic problem for $\tilde{\phi}$ whose solution may be obtained by Fourier transforms, so we define $\phi^* = \int_{-\infty}^{\infty}e^{ikx}\tilde{\phi}(x, z)\,dx$ etc. Equation (F4) reduces to an ordinary differential equation with solution

$$\phi^* = A(k)\exp[-(k^2 - \omega^2/c_2^2)^{\frac{1}{2}}z],$$

where eqn (F6) prescribes the appropriate definition of the square root. Substituting the boundary condition on $z = 0$,

$$A\left[(\omega^2 - c_1^2 k^2)(k^2 - \omega^2/c_2^2)^{\frac{1}{2}} + \frac{\rho_0 w^2}{\sigma}\right] = i\omega F^*. \tag{F7}$$

This leads to an awkward inversion integral for $\tilde{\phi}$ (see Exercise 2.5b) so we adopt a simpler approach, and first look for free waves ($F = 0$) which are periodic in x and decay exponentially as $z \to \infty$, so that

$$\tilde{\phi} = \phi_0 e^{\pm ilx - mz}, \quad (l > 0, m > 0),$$

where from eqn (F4)

$$l^2 - m^2 = \omega^2/c_2^2. \tag{F8}$$

Substituting in eqn (F5) we obtain the relation

$$(c_1^2 l^2 - \omega^2)(l^2 c_2^2 - \omega^2)^{\frac{1}{2}} = \frac{\rho_0 c_2}{\sigma}\,\omega^2 \tag{F9}$$

and the positive square root is to be taken.

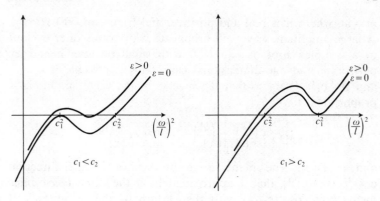

Fig. F1. Roots of eqn (F9)

If we square both sides this may be written

$$G\left(\frac{\omega}{l}\right) = \left[\left(\frac{\omega}{l}\right)^2 - c_2^2\right]\left[\left(\frac{\omega}{l}\right)^2 - c_1^2\right]^2 + c_2^2 \varepsilon^2 \left(\frac{\omega}{l}\right)^4 = 0,$$

where $\varepsilon^2 = \rho_0/\sigma l$ is the fluid damping parameter. The function G is sketched in Fig. F1 for $\varepsilon = 0$ and ε small (>0) in the cases $c_1 < c_2$ and $c_1 > c_2$, and in both cases there is a real root of $G = 0$ with ω/l less than c_1 and c_2. For $c_1 < c_2$ and $\varepsilon \ll 1$ there are two other real roots which are however spurious since they do not satisfy eqn (F9), that is they do not give a decaying signal in the z direction.

For $c_1 > c_2$ there are two complex roots, with small imaginary part if $\varepsilon \ll 1$. For a complex solution to be admissible we require that the waves decay in time and space, and are outgoing. If we take the real part of ω to be positive, then its argument is in $(0, \pi/2)$, so that (ω^2/c_2^2) (c_2 real, positive) has an argument in $(0, \pi)$. Then $l^2 - \omega^2/c_2^2$ has an argument in $(0, -\pi)$ since l^2 is real and positive, and m has an argument in $(0, -\pi/2)$ or $(\pi/2, \pi)$. The condition for an outgoing decaying wave is that m has an argument in $(0, \pi/2)$, and hence these roots are inadmissible. Only one wave can be propagated in the membrane, and its wave speed ω/l will depend on ε and hence on l. We note that if in eqn (F7) we identify k with l, then we obtain eqn (F9) when $F^* = 0$.

However, we are more interested in waves which are forced with a real frequency Ω, and propagate into the fluid so that

$m = in$, where n is real and positive, and from eqn (F8) $\Omega > c_2 l$. A large-amplitude wave will propagate if the value of Ω is close to a complex root of eqn (F9) with small positive imaginary part, as in the one-dimensional case. This is possible if $c_1 > c_2$ and $\Omega = c_1 l + O(\varepsilon)$, so that the inverse of the wave amplitude is proportional to

$$\mathrm{i}\left[c_1^2 - \left(\frac{\Omega}{l}\right)^2\right]\left[\left(\frac{\Omega}{l}\right)^2 - c_2^2\right]^{\frac{1}{2}} + \varepsilon c_2 \left(\frac{\Omega}{l}\right)^2$$

and is $O(\varepsilon)$. The plane wave propagates in the direction $\cos^{-1}[l/(l^2 + n^2)^{\frac{1}{2}}]$, that is approximately at the *coincidence angle* inclined at $\cos^{-1}(c_2/c_1)$ with the surface of the membrane. If $c_2 > c_1$ it is still possible for the wave amplitude to be $O(1/\varepsilon)$ if $\Omega = c_2 l + O(\varepsilon)$, but its direction will make an angle $O(\varepsilon)$ with the x axis and not penetrate significantly into the fluid.

This unique propagation direction may also be obtained from a simple argument in which a point disturbance on the membrane at time $t = 0$ moves along the membrane with speed c_1, and into the fluid radially with speed c_2. If $c_1 > c_2$, then the wavefronts of the disturbance at any time $t > 0$ are as in Fig. F2a and are contained in the wedge of semi-angle $\sin^{-1}(c_2/c_1)$ so that a plane wavefront is propagated in the appropriate direction. Also shown, in Fig. F2b, is a typical plot of the lines of constant magnitude of the wave amplitude for $\Omega = c_1 l + O(\varepsilon)$, $c_1 > c_2$.

The hull sections of the submarine may, however, be more realistically modelled as stiff elastic plates in which the dominant mechanism is the interaction of the bending moment due to the plate curvature. This has already been discussed for a rod in Chapter 1 and the Bernoulli–Euler equation (1.41) derived. For a

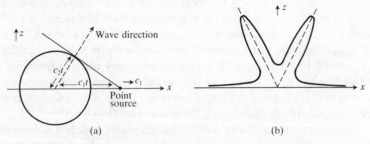

Fig. F2. Wave propagation into a fluid by an elastic membrane

plate of uniform thickness h, surface density γ, with flexure depending on only one space variable x,

$$\sigma_{11} = -x_3 \frac{4\mu(\lambda+\mu)}{\lambda+2\mu} \frac{\partial^2 w}{\partial x^2}, \quad \sigma_{22} = -x_3 \frac{2\lambda\mu}{\lambda+2\mu} \frac{\partial^2 w}{\partial x^2}.$$

This gives rise to a moment about the x_2 axis given by

$$M_2 = -\frac{h^3}{3} \frac{\mu(\lambda+\mu)}{\lambda+2\mu} \frac{\partial^2 w}{\partial x^2}, \tag{F10}$$

and hence resolving in the x_3 direction, as for eqn (1.41),

$$\gamma \frac{\partial^2 w}{\partial t^2} + \frac{h^3}{3} \mu \frac{(\lambda+\mu)}{\lambda+2\mu} \frac{\partial^4 w}{\partial x^4} = -p + \gamma F. \tag{F11}$$

For a more detailed derivation see Graff (1975). A similar discussion to that given for the membrane replaces eqn (F8) by

$$\bar{\mu}^2 l^4 - \omega^2 = \frac{\rho_0 \omega^2}{\gamma i m} \tag{F12}$$

where

$$\bar{\mu}^2 = \frac{h^3}{3\gamma} \frac{\mu(\lambda+\mu)}{\lambda+2\mu} = \frac{Eh^2}{12\rho(1-\sigma^2)}$$

and σ is Poisson's ratio.

A wave of large amplitude will propagate into the fluid if there is a small fluid loading when $\Omega \simeq \bar{\mu}l^2$, provided $\bar{\mu}l > c_2$. The direction of this wave will make an angle $\cos^{-1}(c_2/\bar{\mu}l)$ with the plate surface. There is therefore a striking difference between the membrane and stiff-plate models in that the former propagates no waves into the fluid if $c_1 < c_2$, and sends out a plane wave of large amplitude in a fixed direction if $\Omega \simeq c_1 l$ and $c_1 > c_2$. The latter will always propagate large-amplitude waves of sufficiently short wavelength so that $l > c_2/\bar{\mu}$, but each wavelength propagated will be in a different direction.

The plates are, however, finite in extent, and before drawing too many conclusions from simple solutions an attempt should be made to formulate a problem in which the plate lies in $0 < x < 1$, and in which $w = 0$ outside this region. One approach is to consider a solution of eqn (F4) which corresponds to a (line) source at the origin. In cylindrical symmetry eqn (F4) reduces to eqn (2.26), and an outgoing wave solution is given by the *Hankel function* $H_0^{(1)}(r)$, which is approximately equal to $(2i/\pi)\log r$ for small r (see also Exercise 1.3b). Any distribution of such sources with density f along $z = 0$ will also be a solution of eqn (F4), so we write

$$\phi = \frac{i}{2} \int_{-\infty}^{\infty} f(s) H_0^{(1)} \left\{ \frac{\omega}{c_2} [(x-s)^2 + z^2]^{\frac{1}{2}} \right\} ds. \qquad (F13)$$

Hence

$$\left(\frac{\partial \phi}{\partial z} \right)_{z \to 0} \sim -\frac{1}{\pi} \int_{-\infty}^{\infty} f(s) \frac{z \, ds}{(x-s)^2 + z^2}$$

and $(\partial \phi / \partial z)_{z=0} = f(x)$ as in Exercise 2.5c.

Choose $f(x) = i\omega \tilde{w}$, $(w = \tilde{w}e^{i\omega t})$, so that for the plate problem eqn (F4) becomes

$$\bar{\mu}^2 \frac{\partial^4 \tilde{w}}{\partial x^4} - \omega^2 \tilde{w} = \bar{F} - \frac{i\omega^2 \rho_0}{2\sigma} \int_{-\infty}^{\infty} \tilde{w}(s) H_0^{(1)} \left(\frac{\omega}{c_2} |x-s| \right) ds, \qquad (F14)$$

a singular linear integro-differential equation for $\tilde{w}$. For a finite plate $\tilde{w} = 0$ for $x < 0$ and $x > 1$, so the limits are reduced to $s = 0$ and 1 with $\tilde{w} = 0 = \partial^2 \tilde{w}/\partial x^2$ at the end-points. This problem has to be solved numerically, and the result substituted in eqn (F13) with $f(s) = i\omega \tilde{w}(s)$ to obtain values for ϕ. Further discussion of the solutions of elliptic problems such as eqn (F4) and singular integral equations will be given in Chapter 3.

2.6. Waveguides and dispersion

The wave equation in an infinite domain has the property that any profile can be propagated without change of shape with the wave speed c in a fixed direction. If a boundary is introduced, such general solutions may no longer be possible but *travelling waves* may exist, that is certain specific profiles can be transmitted without change of shape. For example, consider acoustic waves in the strip $0 < y < 1$ where $y = 0$ and 1 are rigid boundaries so that on them $\partial \phi / \partial y = 0$. Because the boundary-value problem is linear with constant coefficients, we look for travelling waves with a complex exponential profile in the form

$$\phi = f(y)e^{i\omega t - ikx}, \quad (\omega > 0),$$

and substitute the expression in eqn (2.30).

The wave speed is ω/lc if a real k can be found satisfying the boundary-value problem

$$f'' + (\omega^2/c^2 - k^2)f = 0; \quad f' = 0 \quad \text{on} \quad y = 0, 1.$$

This has a solution

$$f = A \cos n\pi y, \quad n = 0, 1, 2, \ldots,$$

provided that

$$\omega^2 = c^2(k^2 + n^2\pi^2). \tag{2.34}$$

Then c is the minimum wave speed ω/k and there is a cut-off frequency ck below which no harmonic component propagates. An infinite number of discrete wave speeds are possible for a given k, and the value of ω/k depends on k. The strip $0 < y < 1$ is called a *waveguide*.

In Problem F the boundary $z = 0$ acted as a waveguide and led to a unique value of ω for a given l (the wave number in the x direction corresponding to k, a symbol usually reserved for $(l^2 + m^2)^{\frac{1}{2}}$) and a wave speed ω/l less than c_1 or c_2, which depended on $\varepsilon(l)$.

A more familiar example of a single boundary acting as a waveguide is that of surface gravity waves for an inviscid incompressible fluid. The boundary value problem is

$$\frac{\partial^2 \phi}{\partial x^2} + \frac{\partial^2 \phi}{\partial y^2} = 0, \quad y < 0;$$

$$\frac{\partial \phi}{\partial y} \to 0, \quad y \to -\infty; \quad \frac{\partial^2 \phi}{\partial t^2} + g \frac{\partial \phi}{\partial y} = 0, \quad y = 0.$$

It is easily verified that $\phi = A e^{|k|y + i(\omega t - kx)}$ where

$$\omega^2 = g|k|. \tag{2.35}$$

In this case all frequencies are possible, but again the wave speed ω/k depends on k. Note that in this example a wave is propagated in the fluid in the x direction, which would not be possible without the waveguide action of the surface. In Problem F the waveguide action of the surface modifies the wave which could have been propagated with no surface there.

Waves which propagate with a wave speed which depends on wave number are said to be *dispersive*, and the relation betwen ω

and k is the *dispersion relation*. In two space dimensions the dispersion relation will have the form $\omega = \omega(l, m)$.

Not all waves produced by waveguides are dispersive, and an example is the Rayleigh surface wave in elasticity, which may be used to model earthquake phenomena. We consider linear elastic waves in a two-dimensional space so that the displacement $u = (\phi_x + \psi_y, \phi_y - \psi_x)$, where ϕ and ψ satisfy eqns (2.32) and (2.33) respectively, and we define $c_1^2 = (\lambda + 2\mu)/\rho$, $c_2^2 = \mu/\rho$. If the boundary $y = 0$ is stress-free, then on $y = 0$

$$\sigma_{xy} = \mu\left(2\frac{\partial^2\phi}{\partial x\,\partial y} + \frac{\partial^2\psi}{\partial y^2} - \frac{\partial^2\psi}{\partial x^2}\right) = 0$$

$$= (\lambda + 2\mu)\frac{\partial^2\phi}{\partial y^2} - 2\mu\frac{\partial^2\psi}{\partial x\,\partial y} + \frac{\partial^2\phi}{\partial x^2} = \sigma_{yy}.$$

Hence with

$$\phi = A\,e^{-|\alpha|y + i(\omega t - kx)},$$

$$\psi = B\,e^{-|\beta|y + i(\omega t - kx)},$$

where $\alpha^2 = k^2 - \omega^2/c_1^2$ and $\beta^2 = k^2 - \omega^2/c_2^2$, two homogeneous simultaneous equations may be obtained for A and B. The necessary condition for a non-trivial solution reduces to

$$\left(2 - \frac{c^2}{c_2^2}\right)^2 = 4\left(1 - \frac{c^2}{c_1^2}\right)^{\frac{1}{2}}\left(1 - \frac{c^2}{c_2^2}\right)^{\frac{1}{2}} \tag{2.36}$$

where $\omega/k = c$. This reduces to a cubic in c^2 (ignoring the root $c = 0$) which can be shown to have one real positive root less than c_2 which satisfies eqn (2.36) (see Exercise 2.5a), and has some similarity in structure with eqn (F9). However, this root does not depend on k so that the Rayleigh waves are not dispersive.

For non-dispersive waves a general disturbance will propagate without change of shape, since it may be decomposed into a linear combination of harmonic waves each of which is travelling with the same speed. For dispersive waves only an isolated harmonic wave will travel without change of shape, and a general profile will change its shape as it propagates because each harmonic component is moving with a different speed. Such a profile will in general be given by an expression of the form

$$u = \int_{-\infty}^{\infty} a(k)e^{i[\omega(k)t - kx]}\,dk, \tag{2.37}$$

where we have summed over all possible wave numbers k, with a density function $a(k)$ to be determined by initial conditions. Thus if $u = u_0(x)$ at $t = 0$, where u_0 is complex, then

$$u_0(x) = \int_{-\infty}^{\infty} a(k)e^{-ikx}\,dx,$$

so that $2\pi a(k)$ is the Fourier transform of $u_0(x)$, and

$$u = \frac{1}{2\pi} \int_{-\infty}^{\infty} \left[\int_{-\infty}^{\infty} u_0(s)e^{iks}\,ds \right] e^{i[\omega(k)t - kx]}\,dk. \tag{2.38}$$

This double integral, which can also be obtained by direct use of Fourier transforms, will be difficult to evaluate in general but may be evaluated asymptotically for large x and t. If $x = Gt$ where G is an $O(1)$ constant and $t \gg 1$, then the method of stationary phase, discussed in Chapter 1, Section 1.4(c), shows that the dominant contribution to the integral comes from wave numbers close to k_0, where k_0 is a root of $d\omega(k)/dk = G$. Thus, at large times, the disturbance will appear to have a wavenumber k to an observer moving with speed G, called a *group velocity*. In most examples G may be interpreted as the speed with which the wave energy is propagated. By expanding the integral about $k = k_0$ it may be shown to decay like $[G(k_0)t]^{-\frac{1}{2}}$ for large t, if $x \sim G(k_0)t$; elsewhere it is exponentially small (see Exercise 2.5b).

The dispersion relations considered so far, namely eqns (F9), (F12), (2.34) and (2.35), have all been monotonic in k, with $G = d\omega/dk$ monotonic in k, but this latter situation will not always occur. Thus if surface tension T is introduced into the surface gravity-wave problem, the surface boundary conditions become

$$y = 0; \quad \frac{\partial^2\phi}{\partial t^2} + g\frac{\partial\phi}{\partial y} = \frac{T}{\rho}\frac{\partial^3\phi}{\partial y\,\partial x^2},$$

and the dispersion relation corresponding to eqn (2.35) is

$$\omega^2 = |k|\left(g + \frac{Tk^2}{\rho}\right) \tag{2.39}$$

This has a minimum value for both ω/k and $\omega'(k)$, so that after a long time there will be no waves in the region $x < G_m t$, and in $x > G_m t$ there will be a superposition of two waves, corresponding to the two roots for k of $x = \omega'(k)t$. (See Exercise 2.6.)

Dispersion will also occur in higher-order wave equations without the introduction of a waveguide. Thus for the stiff plate of Problem F, with no fluid loading, the displacement satisfies eqn (F11) to give

$$\frac{\partial^2 w}{\partial t^2} + \bar{\mu}^2 \frac{\partial^4 w}{\partial x^4} = 0.$$

This leads at once to the dispersion relation $\omega = \bar{\mu} k^2$.

Problem G. Coal seam exploration

In many mining situations a coal seam is embedded in rock, and may run for a considerable distance as a uniform two-dimensional channel. The seam may, however, terminate relatively rapidly either by closing up or by a rapid change of depth due to a geological fault. It is valuable to know when working a seam roughly how much of the seam is left for excavation. It is also useful to know if the seam is likely to meet a region which has already been excavated. One simple way of obtaining data about the structure of a seam is to propagate elastic waves down the seam by small explosions, and to record all reflected signals. The difficulty lies in analysing and interpreting this data, and a simple model of the situation is desirable. The crucial physical features are that the wave speeds in the rock are greater than in the coal, and that the seam can act as a waveguide for the transverse elastic waves defined by eqn (2.33). Thus a signal is produced by a transducer which is dominated by displacements in the horizontal direction normal to the direction of the coal seam, as in Fig. G1. Moreover, the seam is assumed to be very wide so that reflections of these waves come only from the end of the seam. Thus we examine the propagation of transverse waves in the x direction down a seam of infinite width, with displacements normal to the channel cross-section; assuming perfect reflection at $x = D$, the profile at $x = 2D$ for the infinite-length seam is equivalent to the reflected signal back at the coalface. If the channel height is $2d$, and the seam is deep enough so that the rock has infinite expanse above and beneath it, then as in Fig. G1 we look for displacements $\boldsymbol{u} = [0, 0, w(x, y, t)]$ satisfying eqn (2.33), where $\mu/\rho = c_2^2$ in the rock and c_1^2 in the coal (assuming $c_2 > c_1$). The

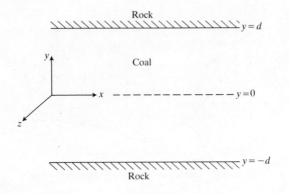

Fig. G1. Wave propagation in a coal seam

conditions at the interface $y = d$ are that the stress and displacements are continuous. Since there is only one stress component σ_{zy}, these conditions reduce to

$$y = d, \quad [w]_1^2 = 0 = \left[\mu \frac{\partial w}{\partial y}\right]_1^2, \tag{G1}$$

where suffix 1 denotes the coal and 2 the rock. For simplicity, look first for symmetrical solutions such that

$$y = 0, \quad \frac{\partial w}{\partial y} = 0; \quad y \to \infty, \quad w \to 0. \tag{G2}$$

Appropriate solutions are called Love waves (Love, 1944) and are given by

$$w_i = f_i(y)e^{i(\omega t - lx)}$$

where

$$f_i'' + f_i\left(\frac{\omega^2}{c_i^2} - l^2\right) = 0, \quad (i = 1, 2).$$

For $i = 2$, $y > d$ we want a decaying solution as $y \to \infty$, so

$$f_2 = A_2 e^{-ny}, \quad c_2^2 n^2 = l^2 c_2^2 - \omega^2 > 0. \tag{G3}$$

For $i = 1$, $0 < y < d$ we want a periodic solution, so

$$f_1 = A_1 \cos my, \quad c_1^2 m^2 = \omega^2 - l^2 c_1^2 > 0. \tag{G4}$$

The inequalities in (G3) and (G4) are satisfied if $c_2^2 > (\omega/l)^2 > c_1^2$.

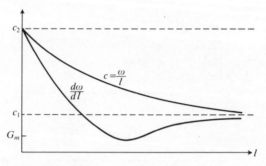

Fig. G2. Dispersion relation for Love waves

If we now substitute the conditions of eqn (G1), the dispersion relation for a travelling wave of this form is obtained in the form

$$\mu_1 m \tan md = \mu_2 n, \tag{G5}$$

where n is the positive root in eqn (G3).

For a given l this has a unique real root in ω, and the wave speed ω/l is plotted in Fig. G2 against l. For small l, $c \simeq c_2$ and

$$(c_2^2 - c^2)\frac{ld}{c_1} \simeq \frac{\mu_2 c_1 2^{\frac{1}{2}}}{\mu_2}(c_2 - c)^{\frac{1}{2}}.$$

For large $\tan md$, $c \simeq c_1$ and

$$kd(c - c_1)^{\frac{1}{2}} 2^{\frac{1}{2}} \simeq \frac{\pi}{2}$$

so that

$$\omega - c_1 l \simeq \frac{\pi^2}{8d^2 l}$$

as $k \to \infty$. Also shown in Fig. G2 is the group velocity, and this has a minimum value $G_m < c_1$ since $d\omega/dl < c_1$ as $k \to \infty$.

Thus if the seam is long so that $D \gg d$, then the signal received at the coal face after a time t_0 will consist mostly of one wave number if $c_1 < 2D/t_0 < c_2$, two wave numbers if $c_1 > 2D/t_0 > G_m$, and no wave at all if $G_m > 2D/t_0$. The most reliable signal will be obtained when $G \simeq G_m$ and $t_0 \simeq 2D/G_m$. Not only should the signal abruptly cut off at later times, but its amplitude should be significantly larger than earlier signals. The reason for this is that the stationary-phase analysis which predicted a signal decay like

$t^{-\frac{1}{2}}$ has to be modified where $d\omega/dl$ has a minimum, and the predicted asymptotic decay rate is then $t^{-\frac{1}{3}}$.

The particular wave investigated is, however, somewhat special, and would require that the transducer was able to produce a signal in the form of a uniform displacement in the z direction which has a prescribed symmetrical form in the y direction. It is not difficult to verify that an antisymmetric transverse wave cannot be found such that the signal decays as $y \to \pm\infty$, and all longitudinal waves will have multiple reflections from the seam walls. Thus if a wave is propagated along the seam it must be a symmetric transverse wave, and any other signals produced by the transducer must be damped out by the multiple reflections.

It is, however, possible to propagate a wave into the rock and not along the channel, just as in Problem F a large-amplitude wave could be propagated into the fluid at the coincidence angle. In this problem we look for $f_2 = A_2 e^{im_2y}$, $f_1 = A_1 e^{imy}$, (with $f_3 = A_3 e^{im_2y}$ in $y < -d$), where $c_2^2(l^2 + m_2^2) = \omega^2$. The boundary conditions are satisfied if $\mu_2 m_2 = \mu_1 m$, so that

$$c_2^2 \mu_1^2 \left[\left(\frac{\omega}{l} \right)^2 - c_1^2 \right] = c_1^2 \mu_2^2 \left[\left(\frac{\omega}{l} \right)^2 - c_2^2 \right].$$

The wave direction is given by

$$\cos\alpha = \frac{c_2 l}{\omega} = \frac{1}{c} \left(\frac{c^2 - \mu^2}{1 - \mu^2} \right)^{\frac{1}{2}},$$

where $\mu = \mu_1/\mu_2$, $c = c_1/c_2$, and exists for $\mu_1/\mu_2 < c_1/c_2 < 1$. Thus for a given frequency ω, waves with wave number (ω/c_1) $[(c^2 - \mu^2)/(1 - \mu^2)]^{\frac{1}{2}}$ will 'leak out' into the rock, whereas those in the range ω/c_2 to ω/c_1 will propagate along the seam. We shall return to this problem again in Chapter 5.

2.7. Quasi-linear systems

We return here to a suffix notation for derivatives. In Section 2.3 the theory of linear hyperbolic systems was discussed, and a quasi-linear first order system is defined by

$$A\boldsymbol{u}_t + B\boldsymbol{u}_x = \boldsymbol{d}, \tag{2.40}$$

where A and B are matrices, and $\boldsymbol{d}$ is a vector, which are functions of u, x, t. In particular, hyperbolic systems for which A and B are functions of u alone and $d = 0$ are of considerable interest, because they necessarily possess Riemann invariants for a system of two equations. Following eqn (2.16), characteristic directions are defined by $dx/dt = \lambda$ where λ is a root of $\det(B - \lambda A) = 0$; and along a real characteristic, as before,

$$z^T A \frac{du}{dt} = z^T d, \qquad (2.41)$$

where z^T is the left eigenvector of $B - \lambda A$.

Thus a system of n equations is hyperbolic if there exist n real roots λ, but since A and B are functions of u this will in general not be known before the solution is obtained, and it may well be a hyperbolic problem in only part of the (x, t) domain. A much more serious difficulty is that, given Cauchy data on a suitable initial curve, the solution may exist only in the neighbourhood of the boundary. This is not so in the linear case.

Discontinuities in u_t and u_x propagate along these characteristics; but, unlike the linear case, discontinuities in u, satisfying a weak form of the equations, will not propagate along the characteristics. It is now much more difficult to define a weak solution, and only when the equations are in conservation form, that is

$$\frac{\partial u}{\partial t} + \frac{\partial}{\partial x} [f(u)] = 0, \qquad (2.42)$$

is it at all obvious how to proceed. In this case, and with Cauchy data on $t = 0$, a weak solution may be defined by a vector function u satisfying

$$\iint_{t>0} [u\psi_t + f(u)\psi_x] \, dx \, dt = -\int_{-\infty}^{\infty} \psi(x, 0)u_0(x) \, dx, \qquad (2.43)$$

where ψ is a differentiable test function which vanishes as $x \to \pm\infty$ and $t \to \infty$. Using Green's Theorem on $t > 0$ eqn (2.43) clearly reduces to eqn (2.42) if ψ is arbitrary. If, however, u is discontinuous across a given curve C, then applying Green's Theorem on both sides of C we obtain

$$\int_C [\psi u] \, dx - \int_C [\psi f(u)] \, dt = 0,$$

and for arbitrary continuous ψ,

$$[u]\frac{dx}{dt} = [f(u)], \qquad (2.44)$$

where [ ] implies the jump discontinuity. This curve is usually called a *shock*, and eqn (2.44) gives the appropriate shock conditions. The weak solution defined in this way is not however unique, and a further condition is necessary to complete the definition. For a discussion of this 'entropy' condition and its generalizations reference may be made to Lax (1973).

A weak solution defined by eqn (2.43) will exist for all $t > 0$, and a classical solution of eqn (2.42) will also satisfy eqn (2.43). A common procedure for constructing a weak solution is to fit together classical solutions with jump conditions across the shocks. These jump conditions have to be given as part of the model in addition to the differential equation, but must be consistent with it, that is they must satisfy a relationship such as eqn (2.44). There may, however, be many possible forms of eqn (2.44), since eqn (2.42) may be derived from many different weak formulations of eqn (2.43). The obvious example is to contrast $u_t + uu_x = 0$ with $uu_t + u^2u_x = 0$, where with $u \neq 0$ the same differential equation leads to different jump discontinuities from eqn (2.44) if in the latter case u is replaced by $u^2/2$.

Problem H. Flow of granular material in a hopper

It is convenient to store coal, which is to be transported, in large hoppers under which a truck can be driven. By opening a shutter in the bottom of the hopper coal flows under gravity into the truck, and when it is full the shutter is closed. A difficulty encountered in this simple procedure is that of 'arching', when with the shutter open the coal particles will not flow and the arch so formed has to be unjammed by mechanical means. This often results in a second difficult flow regime, that of 'coring', where most of the coal is at rest except for a fast moving central core which is difficult to control. Similar phenomena may be observed in a powder flowing in a constricted tube (such as an hour-glass or an egg-timer), and it is clearly valuable to have a model for the flow of a granular material. The shape of the grains will be random, but their size will be roughly the same and much smaller than any length scale relating to the flow geometry, so that a continuous macroscopic model can be attempted. However, unlike classical linear elasticity or Newtonian fluid mechanics we cannot expect the material to possess a simple constitutive law between stress and strain, or rate of strain, because its properties depend on how it is packed. Thus a powder poured into a container may occupy a certain volume, but if the container is gently shaken this volume will pack down, so that the same material has more than one equilibrium state and is quite unlike an elastic solid.

However, conservation of momentum will still apply in any motion, so that part of the model will be

$$\rho \frac{\partial^2 \boldsymbol{u}}{\partial t^2} = \nabla \cdot \underline{\sigma} + \rho \boldsymbol{F} \qquad \text{(H1)}$$

where $\boldsymbol{u}$ is the displacement, ρ the density, $\underline{\sigma}$ the stress tensor and $\boldsymbol{F}$ the body force (gravity). Our first simplification is to consider only slow flow, that is to neglect the inertia terms on the left-hand side of eqn (H1), so that in a two-dimensional situation in non-dimensional variables eqn (H1) becomes

$$\frac{\partial \sigma_1}{\partial x} + \frac{\partial \tau}{\partial y} = 0, \quad \frac{\partial \tau}{\partial x} + \frac{\partial \sigma_2}{\partial y} = 1, \qquad \text{(H2)}$$

where the stress tensor has components

$$\begin{pmatrix} \sigma_1 & \tau \\ \tau & \sigma_2 \end{pmatrix}$$

and is assumed to be symmetric. Thus we have two equations for three unknown components of the stress and need a further equation to complete the model, followed by some relation from which to determine the strain and hence the displacement. For a three-dimensional flow there are six components of stress, and three further relations are required which makes the modelling much more difficult, and this will not be discussed further.

In a two-dimensional flow of a granular material a possible model is to use the Coulomb law of friction, that is to assert that at each point of the material there is a direction along which the material is about to slip. Thus in Fig. H1 consider a direction, the

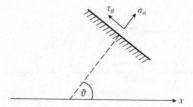

Fig. H1. Stress in polar coordinates

normal to which makes an angle θ with the x axis, where the
normal stress exerted by the material on one side is σ_n and the
tangential stress on shear is τ_θ. If this is the direction of slip and
hence of limiting friction, then $|\tau_\theta| = -\sigma_n \tan \phi$ is an appropriate
generalization of the usual friction law for bodies in contact, since
the normal reaction is $-\sigma_n$. Thus for any orientation at a given
point of the material

$$|\tau_\theta| \leqslant -\sigma_n \tan \phi, \qquad (H3)$$

where equality is attained at one or more orientations θ and ϕ is
the angle of friction, assumed to be a constant of the material. If
τ_θ did not satisfy this inequality then slipping would occur, and an
instantaneous adjustment would take place so that eqn (H3) was
satisfied. To make use of this relation we need to calculate the
stress components at an orientation θ in terms of the principal
stresses σ_1 and σ_2 and shear τ from the tensor relation

$$\begin{pmatrix} \sigma_n & \tau_\theta \\ \tau_\theta & \sigma_\theta \end{pmatrix} = \begin{pmatrix} \cos\theta & \sin\theta \\ \sin\theta & \cos\theta \end{pmatrix} \underline{\sigma} \begin{pmatrix} \cos\theta & -\sin\theta \\ \sin\theta & \cos\theta \end{pmatrix}.$$

This reduces to

$$\sigma_n = -p + \frac{\tau}{\sin 2\psi} \cos 2(\psi - \theta),$$

$$\tau_\theta = \frac{\tau}{\sin 2\psi} \sin 2(\psi - \theta), \qquad (H4)$$

where we have defined

$$p = -\frac{\sigma_1 + \sigma_2}{2}, \quad \tan 2\psi = \frac{2\tau}{\sigma_1 - \sigma_2}. \qquad (H5)$$

Thus p is analogous to a pressure and is never negative, and

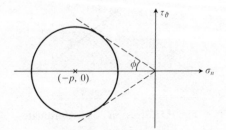

Fig. H2. Mohr circle

directions $\psi = 0$ or $\pi/2$ are directions of zero shear, that is lines of principal stress, with $\psi = 0$ the direction of the smaller principal stress.

Eliminating θ in eqn (H4),

$$\tau_\theta^2 + (\sigma_n + p)^2 = \tau^2 \operatorname{cosec}^2 2\psi,$$

a relation between τ_θ and σ_n which is sketched in Fig. H2 and is called the Mohr circle. Inequality (eqn H3) is indicated by the wedge of semi-angle ϕ, which must touch the circle for equality. Hence from the circle geometry

$$\tau = p \sin \phi \sin 2\psi. \tag{H6}$$

This is the required third relation connecting the three components of stress at each point in the material. From eqns (H5) and (H6)

$$\sigma_{1,2} = -p \pm p \sin \phi \cos 2\psi, \tag{H7}$$

and the problem is reduced to finding two functions p and ψ. Substituting for τ, σ_1 and σ_2 in eqn (H2), these functions satisfy

$$A \begin{pmatrix} p \\ \psi \end{pmatrix}_x + B \begin{pmatrix} p \\ \psi \end{pmatrix}_y = \begin{pmatrix} 0 \\ 1 \end{pmatrix}, \tag{H8}$$

where

$$A = \begin{pmatrix} \sin \phi \cos 2\psi - 1, & -2p \sin \phi \sin 2\psi \\ \sin \phi \sin 2\psi, & 2p \sin \phi \cos 2\psi \end{pmatrix},$$

$$B = \begin{pmatrix} \sin \phi \sin 2\psi, & 2p \sin \phi \cos 2\psi \\ -1 - \sin \phi \cos 2\psi, & 2p \sin \phi \sin 2\psi \end{pmatrix}.$$

Equations (H8) form a quasi-linear system of two equations

which is hyperbolic if $\det(B - \lambda A) = 0$ has two distinct real roots for λ. If we define $\sin \phi = \cos 2\chi$ to simplify this calculation, then $\chi = \pi/4 - \phi/2$, and the roots are given by

$$\lambda = \tan(\psi + \pi/2 \pm \chi). \tag{H9}$$

Thus the characteristics are inclined to the direction of major principal stress at an angle $\pm\chi$, and hence at an angle $\pm(\pi/4 + \phi/2)$ with the minor principal stress direction. It is easily verified from eqns (H4) and (H6) that the slip directions, that is the directions for which eqn (H3) is an equality, are along the characteristics at each point. The left eigenvector z^T of $B - \lambda A$ is easily computed as proportional to $[\cos(\psi \mp \chi), \sin(\psi \mp \chi)]$, and the relationships of eqn (2.41) along the characteristics are then integrable to give Riemann invariants (in the case of negligible gravity) of

$$\pm\cot \phi \log p + 2\psi = \text{constant}. \tag{H10}$$

Hence for problems with Cauchy data given on some initial curve a numerical 'marching' scheme would be effective, and would give a solution valid in some neighbourhood of the initial curve. However, the boundary conditions in the hopper problem are not of this initial-value kind. Thus on the side walls as in Fig. H3 we could assume either that the walls are significantly rougher than the material, so that they are in the direction of slip, or that they are significantly smoother than the material so that they are lines of zero shear, that is the minor principal stress directions at each point. In the first case, the boundary is a characteristic on which a Riemann invariant holds; in the second, ψ is prescribed by the wall geometry.

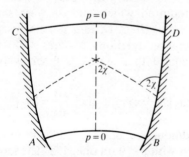

Fig. H3. Granular material in a rough hopper

The arching bottom surface, and possibly the top surface, will be free boundaries, that is there will be zero stress σ_n and shear τ_θ on them, and their position will be unknown. This can only occur if $p = 0$, so that the Riemann invariants along any characteristic intersecting this boundary will be singular, and hence their evaluation in a region close to such a boundary requires more care. One approach is to introduce a small cohesion so that eqn (H3) is replaced by

$$|\tau_\theta| \leqslant -\sigma_n \tan \phi + c,$$

where c is small and positive. Such a term is certainly necessary to describe a soil, but is likely to be very small for a dry powder. The only change is to replace $p \sin \phi$ by $p \sin \phi + c \cos \phi$ in eqns (H6) and (H7). This does not alter the characteristics, but the Riemann invariants become $\pm \cot \phi \log(p + c \cot \phi) + 2\psi$. On a free boundary the zero stress conditions then reduce to

$$\psi = \theta, \quad p = c \cot \chi. \tag{H11}$$

This is not really satisfactory since if BD is a rough wall and hence a characteristic, the Riemann invariant on it will be $O(\log c)$ from conditions at B. Thus we may expect large changes in ψ near to B on BD, or more likely a shock forming so that ψ is discontinuous on some curve meeting BD near B. For a symmetrical situation in a rough hopper, so that both AC and BD are members of the same family of characteristics which contains the line of symmetry, a continuous distribution for p and ψ is also not possible, since characteristics of the second family will inevitably intersect on the centre line at an angle 4χ, less than π.

Thus to describe the stress distribution a weak formulation of the problem is required which allows stress shocks, probably through the corners A, B, C, D, and not necessarily symmetric. From the conservation relations of eqns (H2) it would seem appropriate to require that across a shock whose direction is dy/dx,

$$[\sigma_1]_-^+ \frac{dy}{dx} = [\tau]_-^+; \quad [\tau]_-^+ \frac{dy}{dx} = [\sigma_2]_-^+. \tag{H12}$$

These shock conditions are, however, difficult to use and have no solution consistent with $p = 0$ on one side. The model is therefore inadequate in several respects, but a significantly improved one is not yet available. A more detailed discussion is given by Brennan and Pearce (1978). The case $c \neq 0$, $\phi = 0$ is a model for plasticity and these difficulties do not arise (see Exercise 2.7).

Exercises

2.1. In Problem B, when $1 < c < 3$, obtain the functional equation

$$\dot{P}(t) = a - b\left[8t - 4 + \frac{c}{2}\dot{P}(t) - \frac{\mu c}{2}\dot{P}(\mu t)\right.$$
$$\left. + \frac{c}{2}\dot{P}\left(t - \frac{2}{c+1}\right) - \frac{c}{2\mu}\dot{P}\left(\frac{t}{\mu} - \frac{2}{c-1}\right)\right]$$

valid for times $2/(1+c) < t < 1$. Show that the solution for $\dot{P}(t)$ is a linear function of time, whose coefficients take different values in time intervals defined by $t_n = 1 - \mu^n$. Give a geometrical interpretation in the characteristic diagram for these time intervals, and show that the jumps in $\dot{P}$ and $\ddot{P}$ at $t = t_n$ satisfy

$$[\dot{P}]^{t_n^+}_{t_n^-} = \mu\gamma[\dot{P}]^{t_{(n-1)}^+}_{t_{(n-1)}^-}$$

and

$$[\ddot{P}]^{t_n^+}_{t_n^-} = \gamma[\ddot{P}]^{t_{(n-1)}^+}_{t_{(n-1)}^-}$$

where

$$\gamma = bc/\mu^2(2 + bc).$$

Hence evaluate $\dot{P}$ in $t_n < t < t_{n+1}$, and show that it converges as $t \to 1$ provided $\gamma < 1$.

Briefly indicate how the solution may be obtained for the stiff-spring pantograph and the hanging mass.

2.2. A constant force, moving with speed 1, is applied to a taut wire and creates transverse waves in the wire. If the force is impulsively applied at $x = 0$ at time $t = 0$, evaluate and sketch the wire displacement at a later time t in the cases $c > 1$ and $c < 1$, and explain why the model is invalid when $c \simeq 1$. If the wire has a small flexural rigidity, defined in eqn (1.41), show that in suitably scaled variables the wire slope $u = y_\xi$, where y is the displacement of the wire near the moving point of application of the force, satisfies (when $c = 1$) the equation

$$u_\tau - u_{\xi\xi\xi} = -\delta(\xi),$$

where u vanishes as $\xi \to \pm\infty$ and at $\tau = 0$. Hence obtain an expression for u by using a Fourier transform in ξ. Using the method of stationary phase, show that for large τ and ξ (>0)

$$u \sim \left(\frac{\tau}{\xi}\right)^{\frac{7}{4}} \exp\left(\frac{-2i}{3\sqrt{3}} \frac{\xi^{\frac{3}{2}}}{\tau^{\frac{1}{2}}}\right),$$

and obtain an asymptotic expression for $\xi < 0$. What can one infer from this model about the displacement of the wire when $c = 1$?

Show that, for a force moving with constant acceleration, the slope of the wire displacement is unbounded when the force speed is equal to the wire wave speed, but that this singularity is not propagated along the wire. Contrast this situation with Problem C.

2.3. Show that the Riemann function for the telegraphy equation $\phi_{xy} = \frac{1}{4}\phi$ is $J_0[i(\bar{x} - x)^{\frac{1}{2}}(\bar{y} - y)^{\frac{1}{2}}]$. In Problem D with $\alpha \neq 0$ show that if $\theta_m = T_{2n}(x)$ when $t = nT$, then $T_{2n+1}(x)$ is given by

$$e^{2\alpha x - \tau} T_{2n}(x) + e^{2\alpha x - \tau} \int_0^x T_{2n}(\bar{x}) e^{\alpha(x - \bar{x})} \frac{\partial R}{\partial \bar{x}} (x, t, \bar{x}, 2n\tau) \, d\bar{x}$$

$$- e^{\alpha x} \int_0^\tau e^{(2n+1)(\bar{t} - \tau)} R(x, (2n+1)\tau, 0, 2n\tau + \bar{t}) \, d\bar{t},$$

where $R(x, t, \bar{x}, \bar{t})$ is an appropriate Riemann function. If $\alpha \gg 1$ and $\varepsilon = 0$, by expanding in powers of $1/\alpha$ and ignoring boundary layers on $x = 0$ and 1, show that in an odd cycle

$$\theta_m = T_{2n}(x) - \frac{\bar{t}}{\alpha} T'_{2n}(x) + \frac{\bar{t}(\bar{t} + 2)}{2\alpha^2} T''_{2n}(x) + O\left(\frac{1}{\alpha^3}\right),$$

where $\bar{t} = t - 2n\tau$. Hence argue that a periodic solution is only possible if $T_{2n} = 1 + x(\theta_0 - 1)$ and there are boundary layers of thickness $O(1/\alpha)$ in which θ_g changes by $O(1/\alpha)$ at $x = 0$ and 1.

2.4. An infinite elastic string of density ρ is under tension ρc^2 and initially at rest. Point masses $\lambda \rho l$ are attached at $x = rl$ $(r = 0, \pm 1, \pm 2, \ldots)$, and transverse forces F_b per unit mass act on them for $t > 0$. Assuming a displacement periodic in x, show that the Laplace transform $\hat{s}$ of the transverse displacement s satisfies in $0 < x < l$ the relation

$$\hat{s} = A(p)\cosh\frac{p}{c}\left(x - \frac{l}{2}\right),$$

and, from the slope discontinuity condition at $x = 0$, evaluate A. Homogenize this problem by examining the inversion of $\hat{s}$ as $l \to 0$, showing that it is equivalent to $ct/l \to \infty$. Show that the dominant term is

$$s \sim \frac{\lambda F_b}{(1+\lambda)}\frac{t^2}{2}$$

and that this is equivalent to the homogenized model

$$\frac{\partial^2 s}{\partial t^2} - \bar{c}^2\frac{\partial^2 s}{\partial x^2} = f_b,$$

where $\bar{c}^2 = c^2(1+\lambda)$ and $f_b = \lambda F_b/(1+\lambda)$. (Compare with eqn E5.) If F_b is not constant, when is this procedure likely to be valid?

2.5. (a) Show that for Rayleigh waves in a half space of elastic material, the wave speed c satisfies a cubic in c^2 which has only one real root less than c_2^2 for all values of the elastic constants λ and μ. Show that the two other roots cannot describe a physically possible solution of the equations. If a small periodic normal stress is induced in the surface of the half space, show that it is possible to propagate a wave of large amplitude away from the surface at an angle of $\pi/4$ if λ is small.

(b) Using the stiff-plate model of Problem F, consider the semi-infinite case in which the displacement $w \equiv 0$ in $x < 0$, and show that the Fourier transform ϕ^* of ϕ has the form

$$\phi^* = A(k)e^{-(k^2 - \omega^2/c^2)^{1/2}z},$$

where

$$\left(k^2 - \frac{\omega^2}{c^2}\right)^{\frac{1}{2}}A(k) = -\int_0^\infty w(x, 0)e^{ikx}\,dx = H_+(k)$$

and

$$\left[\left(k^2 - \frac{\omega^2}{c^2}\right)^{\frac{1}{2}}(\omega^2 - \bar{\mu}^2 k^4) + \frac{\rho_0\omega^2}{\sigma}\right]A(k) = i\omega F^* + H_-(k),$$

where H_+ and H_- are analytic functions of k in the upper and lower half planes respectively. This is called a Wiener–Hopf problem (see also Noble (1958)).

(c) By writing $f(s) = f(x) - [f(x) - f(s)]$, show that

$$-\frac{1}{\pi} \int_{-\infty}^{\infty} \frac{z f(s)\, ds}{(x-s)^2 + z^2} \sim f(x)$$

as $z \to 0$, and suggest a possible representation for the delta function $\delta(x)$.

2.6. For infinitesimal simple-harmonic waves propagating over water of mean depth h in a long rectangular channel, show that their phase velocity c and wave number $k = 2\pi/(\text{wavelength})$ are related by the equation

$$c^2 = (g + \gamma k^2) \frac{\tanh kh}{k}$$

in which γ is the coefficient of surface tension divided by the density of water.

Assuming that $h \gg (3\gamma/g)^{\frac{1}{2}}$ but explaining the significance of this assumption, account for the fact that when a slender obstacle is placed across the surface of water which is flowing along a channel with a velocity U satisfying $2(g\gamma)^{\frac{1}{2}} < U^2 < gh$, then a train of waves is developed downstream and another wave-train with smaller wavelength is developed upstream.

2.7. (a) If we model the conditions near a corner of a hopper in Problem H by a rough wall at angle θ_0 to the horizontal in $y > 0$ and a smooth vertical wall in $y < 0$, and neglect both inertia and gravity, show that stress conditions adjacent to the rough wall, determined by $p = p_1$ and $\psi = \psi_1$, must satisfy $\theta_0 = \psi_1 \pm (\pi/4 + \phi/2)$, where ϕ is the friction angle. Assuming that the positive characteristics form a centred fan through the origin, show that in $y < x \tan(\pi/4 - \phi/2)$

$$p = p_1 \exp[(-2\theta_0 + 3\pi/2 + \phi)\tan \phi].$$

(b) If the rough wall in the previous case feeds into a smooth horizontal wall and the normal stresses are constant in the horizontal section, show that their magnitudes are $p_0(1 \pm \sin \phi)$, where p_0 is a root of

$$(p_0^2 + p_1^2)\cos^2\phi = 2p_0 p_1[1 + \sin^2\phi \, \sin(2\theta_0 - \phi)],$$

if the jump conditions of eqn (H12) are used.

(c) A model for plasticity is to assume a zero friction angle in the granular-material stress model with a non-zero cohesion c; gravity is also neglected. Show that the model has Riemann invariants $p \pm 2c\psi$, where the characteristic directions bisect the directions of principal stress. Obtain a solution for the stress field near a boundary with a corner.

3 Elliptic problems

3.1. Potential theory

In Chapter 2 a linear partial differential equation in two independent variables was defined to be hyperbolic if there exist two real distinct characteristic directions at each point, which define curves across which discontinuities in the first derivatives of the unknown function ϕ can occur. For an *elliptic* equation no real characteristic directions exist, and eqn (2.6) has no real roots for dy/dx. Hence $B^2 < AC$ and the integral solutions of eqn (2.6) have the form $p(x, y) \pm iq(x, y) = \text{constant}$, where q is not identically zero. A transformation to new variables p and q reduces the general second-order equation (2.5) to the canonical form

$$\phi_{pp} + \phi_{qq} + d\phi_p + e\phi_q + f\phi = g, \qquad (3.1)$$

where d, e, f and g are functions of p and q. The highest-order derivatives appear in the form $\nabla^2\phi$ where ∇^2 is the Laplacian operator with respect to p and q. The non-existence of characteristic curves implies that the properties of elliptic equations are very different from hyperbolic ones, and less explicit analytical information can be obtained. Hyperbolic equations with given initial data form well-posed problems, and as such are analogous to ordinary differential equations with given initial data. Elliptic equations are analogous to two-point boundary problems for ordinary differential equations, and the technique most commonly used to develop their qualitative properties is that of the Green's function, suitably generalized from its definition in Chapter 1.

For the linear elliptic equation in canonical form in x, y variables we define a Green's function $G(x, y; \xi, \eta)$ by

$$G_{xx} + G_{yy} + dG_x + eG_y + fG = \delta(x - \xi)\,\delta(y - \eta), \qquad (3.2)$$

together with a homogeneous boundary condition on a closed curve C. Note that an alternative definition following Exercise

1.5 is

$$G_{\xi\xi} + G_{\eta\eta} - (dG)_\xi - (eG)_\eta + fG = \delta(x - \xi)\,\delta(y - \eta), \quad (3.3)$$

and that G will not be symmetric in x, y and ξ, η unless the differential operator is self-adjoint, that is $d = e = 0$. To avoid a number of technical difficulties we consider only self-adjoint problems, and first discuss the case $f = 0$ when eqn (3.1) reduces to Poisson's equation. If we also require that $\phi = h$ on the closed curve C, and hence choose $G = 0$ on C, then by considering the Green's identity

$$\iint_D [G(\phi_{\xi\xi} + \phi_{\eta\eta}) - \phi(G_{\xi\xi} + G_{\eta\eta})]\,d\xi\,d\eta = \int_C \left(G\frac{\partial\phi}{\partial n} - \phi\frac{\partial G}{\partial n}\right)d\sigma$$

over the domain D enclosed by C,

$$\phi(x, y) = \iint_D Gg\,d\xi\,d\eta + \int_C h\frac{\partial G}{\partial n}\,d\sigma. \quad (3.4)$$

Thus if the Green's function is known, the solution is given by quadratures from eqn (3.4) and we have solved the Dirichlet problem. The Green's function will depend on the closed curve C (and on the differential operator in a more general case), but is independent of the inhomogeneous term g in eqn (3.1) and the boundary data h. It is difficult to explicitly calculate G except for simple domains D such as a circle (or a half space), but their existence and uniqueness is assured provided that there is no non-trivial solution to the problem for ϕ when $g \equiv 0$ and homogeneous boundary conditions are posed (see Exercise 3.1a). For the Neumann problem $\partial\phi/\partial n = h$ on C, there is a trivial solution to the homogeneous problem, which is a constant; and in general G, and hence the solution ϕ, does not exist. This is easily verified since from Green's theorem $\iint \nabla^2\phi\,d\xi\,d\eta = \int (\partial\phi/\partial\nu)\,d\sigma$, and hence for a solution to exist $\iint g\,d\xi\,d\eta = \int h\,d\sigma$. If this is satisfied then G and ϕ exist, but are only unique up to an additive constant. This is a further example of the Fredholm alternative described in Chapter 1 with reference to Fredholm integral equations.

It is not difficult to extend the result of eqn (3.4) to the case $f \neq 0$, called Helmholz's equation, when $f \geq 0$ in D; but there will now be non-trivial solutions of the homogeneous Dirichlet problem for ϕ, that is eigensolutions, for an infinite set or a discrete range of positive values of f. These eigenvalues will depend on the boundary shape C, and in general cannot be expressed in a simple form. Thus the determination of the existence and uniqueness of a solution of Helmholz's equation will first require the calculation of the eigenvalues associated with the given boundary shape. Nor is it difficult to extend the result for Poisson's equation to three dimensions; eqns (3.2) and (3.4) remain unaltered in form except for the additional space dimension.

Poisson's equation, or Laplace's equation when $g \equiv 0$, occurs as a simple model in a variety of physical situations. For inviscid irrotational incompressible fluid motion div $\boldsymbol{q} = 0 = $ curl $\boldsymbol{q}$, and the velocity potential, defined by $\phi = \int \boldsymbol{q} \cdot d\boldsymbol{x}$ so that $\boldsymbol{q} = $ grad ϕ, satisfies $\nabla^2 \phi = 0$. For the theory of gravitation, where ϕ is the gravitational potential of a distribution of matter of density $\rho > 0$ and γ is the gravitational constant, $\nabla^2 \phi + \gamma \rho = 0$. Similarly for electrostatics with a charge distribution ρ in suitable units $\nabla^2 \phi + \rho = 0$, where ϕ is the electrostatic potential such that the field $\boldsymbol{E} = -$grad ϕ. Thus the Green's function for the Dirichlet problem may be interpreted as the electrostatic potential due to a point charge (in three dimensions) at the point (ξ, η, ζ) in the presence of an earthed conductor. For the whole space this is clearly $G = 1/4\pi r$, where $r^2 = (x - \xi)^2 + (y - \eta)^2 + (z - \zeta)^2$, called the *fundamental solution* of Laplace's equation in three dimensions. In two dimensions the fundamental solution is $-(1/2\pi)\log r$, where $r^2 = (x - \xi)^2 + (y - \eta)^2$, and is the Green's function for the whole space (see Exercise 3.1b).

To obtain closed-form solutions, only three techniques are available: (i) to separate the variables for a simple enough geometry; (ii) to take a Fourier or associated transform suitable for the geometry of the problem; and (iii) (in two dimensions only) to use the methods of complex variable theory.

Problem I. Current flow between two plates joined by spigots

A large-scale industrial device consists of two equal rectangular horizontal plates, sides a and ab, separated by a number of thin hollow columns called spigots, as in Fig. I1. The plates are thin and are situated distance h apart, one vertically above the other.

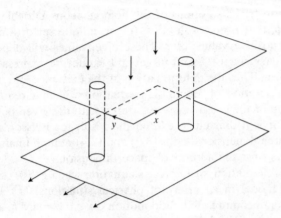

Fig. I1. Current flow between two plates joined by spigots

The spigots are all identical, with thin walls and a small circular cross-section. Electric current is fed into the top plate with a constant volume density, flows down the spigots into the lower plate, and out of the system through one edge of the lower plate. The edges of the upper plate and the other three edges of the lower plate are insulated. For spigots of a given radius ca, p spigots are needed to mechanically support the top plate. The design problem is to distribute the spigots in such a way that the energy loss from the system in a steady state is minimized. If this can be done then the optimum number of spigots may be determined, given that the spigot radius will vary inversely with p.

The first step is to determine the electric potential Φ at all points of the system, where the current flow is $\boldsymbol{j} = -\sigma \nabla \Phi$ and $\nabla^2 \Phi = -$(source density of current). Here σ is the conductivity per unit area of the plates assumed constant in each plate. To take advantage of the small cross-section of the spigots we treat them as wires, carrying total currents i_s at $\boldsymbol{x} = \boldsymbol{x}_s = (x_s, y_s)$ with $s = 1, 2, \ldots p$, in determining the potentials in the upper and lower plates. Then a suitably normalized potential $\phi(x, y)$ for the upper plate satisfies

$$\nabla^2 \phi = -1 + \sum_1^p i_s \delta(\boldsymbol{x} - \boldsymbol{x}_s), \tag{I1}$$

where $\delta(\boldsymbol{x})$ is the two-dimensional delta function, equal to zero if

$x \neq 0$, but with

$$\iint_D \delta(x) \, \mathrm{d}x \, \mathrm{d}y = 1$$

if D contains $x = 0$. Boundary conditions at the edge of the plate are

$$\phi_y = 0 \text{ on } y = 0 \text{ and } b; \quad \phi_x = 0 \text{ on } x = 0 \text{ and } 1, \qquad (I2)$$

where distances have been scaled with a. The total current $b = \sum_1^p i_s$, but the individual values of i_s are to be determined.

For the lower plate, with the potential ψ normalized as for ϕ,

$$\nabla^2 \psi = -k \sum_1^p i_s \delta(x - x_s), \qquad (I3)$$

where k is the ratio of the two-dimensional conductivities in the upper and lower plates respectively, and if made of the same material is the ratio of their thicknesses. In addition

$$\psi_y = 0 \text{ on } y = 0 \text{ and } b; \quad \psi_x = 0 \text{ on } x = 1; \quad \psi = 0 \text{ on } x = 0;$$

and from conservation of total current

$$\int_0^b (\psi_x)_{x=0} \, \mathrm{d}y = kb.$$

The two plates are linked by conditions along each spigot, and we assume that for both plates each spigot is equivalent to a circular hole in the plate, of radius c and centre x_s, whose boundary is an equipotential and from which current is either removed or introduced. For this to be a reasonable approximation c must be small compared to distances between the spigots, but not so small that the potential ϕ_s on the boundary of the hole is unacceptably large. Thus for the top plate, current is removed with density i_s at an equipotential ϕ_s where, although $\phi \simeq (i_s/2\pi)\log|x - x_s|$ as $x \to x_s$, $\phi_s \neq (i_s/2\pi)\log c$; that is, $\log c$ is not too large compared to unity.

For the bottom plate, similar expressions hold with $\psi \simeq -(ki_s/2\pi)\log|x - x_s|$ as $x \to x_s$, and for simplicity we only consider the case in which $k \ll 1$, that is the bottom plate is much thicker than the top plate. This implies from eqn (I3) that $\psi = O(k)$ and is zero in our approximation, so that the bottom end of each spigot has zero potential. If each spigot has resistance R_s, then to complete the boundary conditions for the top plate $\phi_s = i_s R_s$.

The boundary-value problem is to solve eqns (I1) and (I2) in the top plate with unknown constants i_s (such that $b = \sum_s i_s$), evaluate the solution on each spigot to obtain ϕ_s, and hence determine i_s from a system of simultaneous equations. Since the solution will only give ϕ as approximately constant on each spigot cross-section, ϕ_s is determined from the mean value. The energy dissipated in the system can then be calculated from

$$E = \tfrac{1}{2} \int\!\!\int (\nabla\phi)^2 \, \mathrm{d}x \, \mathrm{d}y + \sum_1^p R_s i_s^2,$$

where the domain of integration is the top plate less the spigot holes. Using Green's Theorem this simplifies to

$$E = -\tfrac{1}{2} \int\!\!\int \phi \nabla^2 \phi \, \mathrm{d}x \, \mathrm{d}y + \int_{\text{spigots}} \phi \frac{\partial \phi}{\partial n} \, \mathrm{d}s + \sum_1^p R_s i_s^2 = \tfrac{1}{2} \int\!\!\int \phi \, \mathrm{d}x \, \mathrm{d}y;$$

$$\text{(I4)}$$

where we have used eqn (I1) and the conditions on each spigot, and the domain of integration is the upper plate less the spigot cross-sections. Finally some optimization procedure must be developed to determine the best location of the spigot holes. For values of p greater than 2 or 3 this will inevitably imply a numerical procedure, both for solving the boundary-value problem and the consequent optimization. To give some guidance and a check on any such procedure we attempt to solve the boundary-value problem by transforms in the case $p = 2$ (Fig. I2).

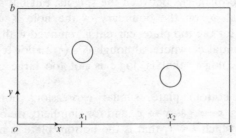

Fig. I2. Boundary value problem for top plate

This is a bounded domain, so we use the finite Fourier transform introduced in Exercise 1.4 and define

$$\phi_n^* = \int_0^b \phi(x, y)\cos\frac{n\pi y}{b}\,\mathrm{d}y, \quad n = 0, 1, \ldots,$$

which has an inversion formula, obtained from the Fourier Series Theorem, of

$$\phi(x, y) = \frac{1}{b}\left(\phi_0^* + 2\sum_1^\infty \phi_n^*\cos\frac{n\pi y}{b}\right). \tag{I5}$$

Then eqn (I1) becomes

$$\frac{\mathrm{d}^2\phi_0^*}{\mathrm{d}x^2} = -b + \sum_1^2 i_s\delta(x - x_s),$$

$$\frac{\mathrm{d}^2\phi_n^*}{\mathrm{d}x^2} - \frac{n^2\pi^2}{b^2}\phi_n^* = \sum_1^2 i_s\cos\frac{n\pi y_s}{b}\,\delta(x - x_s), \quad n > 0,$$

with boundary conditions $\mathrm{d}\phi_n^*/\mathrm{d}x = 0$ at $x = 0$ and 1 for all n. Integrating,

$$\phi_0^* - C = \frac{-bx^2}{2}, \quad 0 < x < x_1,$$

$$= \frac{-bx^2}{2} + i_1(x - x_1), \quad x_1 < x < x_2, \tag{I6}$$

$$= \frac{-bx^2}{2} + i_1(x - x_1) + i_2(x - x_2), \quad x_2 < x < 1,$$

where C is an undetermined constant and $i_1 + i_2 = b$.

Note that this is yet another example of the Fredholm alternative in which there is a constant solution to the homogeneous problem, and hence a necessary condition on the inhomogeneous term for there to be a solution, which will not then be unique. To solve ϕ_n^* we need a Green's function defined by

$$\frac{\mathrm{d}^2 G_n}{\mathrm{d}x^2} - \frac{n^2\pi^2}{b^2}G_n = \delta(x - \xi); \quad \frac{\mathrm{d}G_n}{\mathrm{d}x} = 0, \quad x = 0 \text{ and } 1.$$

Such a function may be constructed in the form

$$
G(x, \xi) = \frac{-b}{n\pi} \frac{\cosh \dfrac{n\pi x}{b} \cosh \dfrac{n\pi(1-\xi)}{b}}{\sinh \dfrac{n\pi}{b}}, \quad 0 < x < \xi,
$$

$$
= \frac{-b}{n\pi} \frac{\cosh \dfrac{n\pi(1-x)}{b} \cosh \dfrac{n\pi\xi}{b}}{\sinh \dfrac{n\pi}{b}}, \quad \xi < x < 1. \quad (I7)
$$

It is easily verified that this satisfies the boundary conditions and equation, except at $x = \xi$, where G is continuous but dG/dx is discontinuous. G is also symmetric because the equation is self-adjoint. Hence we may write

$$
\phi_n^* = \sum_1^2 i_s \int_0^1 \cos \frac{n\pi y_s}{b} \delta(\xi - x_s) G_n(x, \xi) \, d\xi
$$

$$
= \sum_1^2 i_s \cos \frac{n\pi y_s}{b} G_n(x, x_s). \quad (I8)
$$

From the integral of eqns (I6) and (I7), on substituting into eqn (I5) we obtain ϕ, which can be evaluated on two spigots to give two equations for the unknown constant in the integral of eqn (I6) and for the currents, which are related by $b = i_1 + i_2$. The evaluation of the coefficients in these equations may, however, be simplified by truncating ϕ after one or two terms, since the coefficients decrease rapidly with n (see Exercise 3.1c).

3.2. Variational methods and variational inequalities

Many elliptic boundary-value problems may be reformulated in terms of a variational problem; that is, the function ϕ which satisfies the boundary-value problem may also be associated with the minimum value of a variational integral. Thus we consider

$$
E(\psi) = \iint_D [\tfrac{1}{2} |\nabla \psi|^2 - f(\psi)] \, dx \, dy, \quad (3.5)
$$

where D is a fixed domain and ψ is a differentiable function which takes prescribed values on its boundary ∂D. Then for a small variation η, where $\eta = 0$ on ∂D, the use of Green's Theorem gives

$$E(\psi + \eta) - E(\psi) = \iint_D [\psi_x \eta_x + \psi_y \eta_y - \eta f'(\psi)]\, dx\, dy + O(\eta^2)$$

$$= -\iint_D \eta [\psi_{xx} + \psi_{yy} + f'(\psi)]\, dx\, dy + O(\eta^2).$$

If η is arbitrary, except for appropriate continuity properties (see Hildebrand (1975) for details), the term δE linear in η will vanish when $\psi = \phi$, where ϕ satisfies

$$\phi_{xx} + \phi_{yy} + f'(\phi) = 0 \quad \text{in } D, \tag{3.6}$$

called the Euler equation for the variational integral of eqn (3.5). This is the condition for $E(\psi)$ to have a stationary value at $\psi = \phi$ and it can usually be demonstrated that it is a minimum, so that an alternative statement is that if ϕ satisfies eqn (3.6), then $E(\phi) \leqslant E(\psi)$ for all admissible functions ψ. Every variational integral has an associated Euler equation, that is the extremal (or minimizer) ϕ satisfies a differential equation; but the converse is not true, and there are many differential equations not equivalent to the Euler equation of a variational integral, that is to a variational principle. In physical applications a variational principle is usually in some sense a conservation law or minimum energy statement, so that for a dissipative system we would not necessarily expect a variational principle to exist.

A number of powerful numerical methods have been developed for solving variational problems, including the classical Rayleigh–Ritz procedure described in Hildebrand (1975).

A variational formulation is particularly appropriate for problems in which there are constraints on the solution over all or part of the domain D or its boundary ∂D. To illustrate this consider the *obstacle problem* for an elastic membrane, that is we attempt to stretch a membrane over a irregular surface so that contact is only achieved in parts of D as in Fig. 3.1. If q is the pressure exerted on the upper side of the membrane then the displacement w satisfies (in steady conditions)

$$\nabla^2 w + q = 0,$$

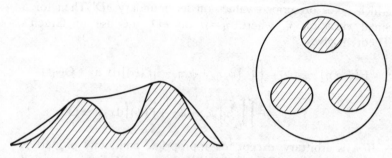

Fig. 3.1. Membrane stretched over an obstacle

where there is no contact with the surface, that is in $D - \Omega$. In the region Ω, w is prescribed; it is also prescribed on ∂D. We also require that ∇w is continuous across the boundary of Ω, which is of course unknown. This leads to a free-boundary problem on the domain $D - \Omega$, but we attempt to construct a variational statement on the whole domain D. Thus we have to solve the constrained problem

$$-\nabla^2 w \geqslant q, \quad w \geqslant g \quad \text{in } D, \tag{3.7}$$

where $z = g(x, y)$ is the prescribed obstacle; the first inequality arises because where there is contact the surface exerts a positive pressure. This problem may be reformulated as a variational inequality, that is $E(w) \leqslant E(v)$, provided that the set of admissible functions is more carefully specified.

The general theorem is discussed in Elliott and Ockendon (1982), but for our applications we require that the functions v have the following properties:

  (i)   they assume prescribed values on ∂D;
  (ii) they have continuous bounded first derivatives and satisfy $v \geqslant g$ in D; and
  (iii) they belong to a *convex* set, that is given two such functions v_1 and v_2, then $\lambda v_1 + (1 - \lambda) v_2$ is also a member of the set for all λ in $0 \leqslant \lambda \leqslant 1$.

It may then be shown that there exists a unique function w such that for $q \geqslant 0$,

$$\iint_D \nabla w \cdot \nabla (v - w) \, dx \, dy \geqslant \iint_D q(v - w) \, dx \, dy \tag{3.8}$$

for *all* values of v.

For the membrane problem

$$E(v) = \int \int \left[\tfrac{1}{2} |\nabla v|^2 - qv \right] dx \, dy,$$

which is the elastic energy functional, and it may be verified that $E(w) \leq E(v)$ using eqn (3.7). This then reduces to eqn (3.8), which gives a reformulation of the problem in terms of a variational inequality. Numerical methods for solving variational inequalities are described in Elliott and Ockendon (1982).

Problem J. *Electrochemical painting*

A convenient method for painting metal objects is by immersing the 'workpiece' in an electrolyte solution containing large negatively-charged organic ions. The ions are neutralized by imposing a potential difference between the workpiece as anode and the earthed bath containing the electrolyte, as in Fig. J1. The applied voltage induces a number of electrochemical and transport processes which result in paint being deposited on the workpiece. However, not all the immersed surfaces of the workpiece are necessarily painted and it is important to determine the 'paint throw', that is the part of the surface that is covered, and the thickness of the paint film which is deposited. Typically a re-entrant corner as shown in Fig. J1 may fail to be painted, and a simple model is needed which can describe this phenomenon.

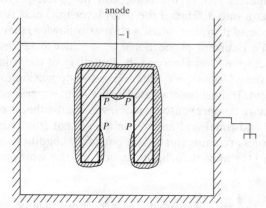

Fig. J1. Workpiece immersed in an electrolyte solution

The process is complicated, but we focus on conditions which hold a few seconds after a potential difference of unity is applied, when there appears to be effectively a steady state in which the paint thickness and paint throw change very slowly. We assume that the concentration of paint ions in the bath is constant (both in space and time due to stirring), and that gross charge neutrality is maintained. Thus if $\phi(x) - 1$ is the electric potential in the electrolyte, then a steady current will be conducted such that $j = -\sigma \nabla \phi$, where div $j = 0$ and

$$\nabla^2 \phi = 0. \tag{J1}$$

This may also hold inside the paint layer adjacent to the workpiece, but the conductivity $\bar{\sigma}$ of the paint will be much smaller than σ, possibly $\bar{\sigma}/\sigma \simeq 10^{-6}$. Thus, from continuity of current, the potential gradient across the paint layer will be large and approximately constant, and the thickness of the paint layer h will be small. If $\phi_s - 1$ is the potential and j_s the normal current flow at the surface between the paint layer and the electrolyte, then

$$h \simeq \frac{\bar{\sigma}\phi_s}{j_s} = \frac{\bar{\sigma}}{\sigma} \frac{\phi_s}{\left(\dfrac{\partial \phi}{\partial n}\right)_s}. \tag{J2}$$

With no current flowing a dissolution of paint is observed if a painted film is left in the bath, and the paint returns to the solution at a rate d. Thus if the paint layer thickness is to remain roughly constant in time when a current is flowing, there must be a dissolution current proportional to d which balances the outflow of negative ions. Hence at the boundary of the paint layer in a steady state $(\partial \phi / \partial n)_s = \varepsilon$, where ε is the dimensionless dissolution current. It must be evaluated either from experiment or from a much more comprehensive model; we assume that it is constant (in space and time) and not too small, so that from eqn (J2) the film thickness remains small. The boundary conditions associated with eqn (J1) may then be evaluated on the workpiece in the form

$$\frac{\partial \phi}{\partial n} = \varepsilon \quad \text{if painted;} \quad \phi = 0 \quad \text{if not painted.} \tag{J3}$$

Moreover we have the obvious constraints

$$\phi > 0 \quad \text{if painted;} \quad \frac{\partial \phi}{\partial n} < \varepsilon \quad \text{if not painted.} \qquad (J4)$$

On the bath $\phi = 1$, and in the bath $\phi > 0$.

These are somewhat unusual boundary conditions, since the points P at which the alternative conditions of eqn (J3) switch is unknown, and their determination is the most interesting part of the problem. It is not immediately clear that there is a unique solution to the problem as posed, but it will turn out to be well posed if we require that $\nabla \phi$ is bounded at P, a very reasonable physical condition. In fact the boundary-value problem may be written as a variational inequality in the form

$$\int_D \nabla \phi \cdot \nabla (v - \phi) \, \mathrm{d}\tau + \varepsilon \int_{\text{workpiece}} (v - \phi) \, \mathrm{d}S \geq 0, \qquad (J5)$$

where the admissible functions v have the following properties:
   (i) $v = 1$ on the bath, $v \geq 0$ on the workpiece; and
   (ii) they have continuous bounded first derivatives and belong to a convex set.
This equivalence is easily verified as in Aitchison, Lacey and Shillor (1984), where numerical methods for the solution of eqn (J5) are described and inadequacies of the model discussed.

3.3. Free-boundary problems

In Chapter 1 the classical model of eqn (1.29) for water waves was established; it consists of an elliptic equation $\nabla^2 \phi = 0$, bounded in part by an unknown boundary on which two non-linear boundary conditions are given, derived from conservation of mass and momentum at the boundary. These two conditions, rather than the single condition required for a fixed boundary-value problem, are sufficient in general to determine the boundary.

Free-boundary problems of this kind also occur in models for slow viscous flow between two parallel plates a small distance ε apart, called a Hele–Shaw cell. The equations of slow flow

$(Re \ll 1)$ are, from eqn (1.24),

$$\nabla^2 q = \frac{1}{\mu}\nabla p, \quad \nabla \cdot q = 0.$$

We rescale distances and velocities normal to the plates (the z direction) with ε, rescale the pressure with $1/\varepsilon^2$ and ignore $O(\varepsilon)$ terms to obtain

$$\frac{1}{\mu}\bar{p}_x - u_{\bar{z}\bar{z}} = 0 = \frac{1}{\mu}\bar{p}_y - v_{\bar{z}\bar{z}}, \quad u_x + v_y + \bar{w}_{\bar{z}} = 0 = \bar{p}_{\bar{z}}.$$

Integrating with respect to $\bar{z}$ using $u = v = 0$ on $\bar{z} = 0$ and 1,

$$u = \frac{1}{2\mu}\bar{p}_x\bar{z}(\bar{z}-1), \quad v = \frac{1}{2\mu}\bar{p}_y\bar{z}(\bar{z}-1), \tag{3.9}$$

where $\bar{p}_x$ and $\bar{p}_y$ are independent of $\bar{z}$.

Integrating for $\bar{w}$ and using $\bar{w} = 0$ on $\bar{z} = 0$ and 1,

$$\nabla^2\bar{p} = \bar{p}_{xx} + \bar{p}_{yy} = 0. \tag{3.10}$$

For a fluid or free boundary of a Hele–Shaw cell, given by $f(x, y, t) = 0$, conservation of momentum normal to the boundary gives, on average across the layer, $\bar{p} = \text{constant}$ if surface tension is constant; and, since $\bar{p}$ is large after scaling with $1/\varepsilon^2$, this constant is zero. The conservation of mass condition is that

$$\frac{df}{dt} = 0 = \frac{\partial f}{\partial t} + q \cdot \nabla f,$$

where q is the average velocity across the layer, equal to $-(1/12\mu)\nabla\bar{p}$ from eqn (3.9). Hence the second condition is $\partial f/\partial t = (1/12\mu)\nabla\bar{p} \cdot \nabla f$. Since $f = 0$ when $\bar{p} = 0$ for all t, $f = \bar{p}$ on the boundary, and the final form of the boundary condition can be written as

$$f = \bar{p} = 0, \quad \frac{\partial\bar{p}}{\partial t} = \frac{1}{12\mu}|\nabla\bar{p}|^2. \tag{3.11}$$

A third model of this kind describes the flow or percolation of a liquid in a saturated porous medium. The fluid velocity q and pressure p are related by the Darcy law $q = -k\,\text{grad}(p + \rho gz)$, where k is a positive constant called the *permeability*. This law, well tested by experiment, takes the place of a momentum balance and constitutive relation, and may be derived by a homogenization procedure (see Neuman (1977)) from equations for the flow of a viscous fluid in a porous matrix. Applying conservation of mass, and assuming that k is constant, we have $\nabla^2 p = 0$, The conditions on a free or fluid boundary $f(xyzt) = 0$ are then

$$p = \text{constant}, \quad \frac{\partial f}{\partial t} = k \ \text{grad}(p + \rho g z) \cdot \nabla f. \qquad (3.12)$$

Equations (3.11) and (3.12) demonstrate the close analogy between the Hele–Shaw cell and flow in a porous media.

We have seen in the preceding section that the obstacle problem could be considered as a free-boundary problem, but that it was also equivalent to a variational inequality. This equivalence will not always be available, although the use of a Baiocchi transformation as in Exercise 3.2 is effective for some problems with boundary conditions of the form of eqns (3.11) or (3.12). For steady flows in two dimensions with no gravity term, the techniques of complex variable theory and conformal mapping may be used to derive a wide variety of solutions as in Birkhoff and Zarantanello (1957). With gravity retained there are a few problems which can be solved by very special transformations as in Polubarinova-Kochina (1962).

A class of problems which can be tackled analytically are those flows which are small perturbations about some exact solution, such as a state of rest or a uniform motion. The problem may then be reduced to a linear boundary-value problem on a fixed domain, and solved by the methods of the previous sections. The most well-known example is that of linearized water-wave theory (or Stokes waves) as in eqn (1.30), where Fourier transforms are widely used to give useful results. This is a linearization about a state of rest, or uniform motion parallel to the free boundary, which is clearly stable to small disturbances. Many free boundaries may not, however, be stable to small disturbances, and the small-perturbation procedure may be used to discuss their stability. Consider, for example, inviscid liquid of mean depth h contained in a cylindrical tank, whose axis is in the vertical direction and which is oscillating up and down as in Fig. 3.2.

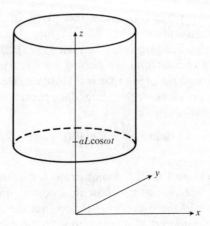

Fig. 3.2. Vertical oscillations of a cylindrical tank of fluid

The boundary conditions on the fluid are

$$\frac{\partial \phi}{\partial n} = 0 \text{ on the cross-section boundary,}$$

$$\frac{\partial \phi}{\partial z} = -\alpha L \omega \sin \omega t \quad \text{on } z = \alpha L \cos \omega t, \text{ the bottom of the tank,}$$

$$\frac{\partial \phi}{\partial z} = \frac{\partial \eta}{\partial t} + \frac{\partial \phi}{\partial x}\frac{\partial \eta}{\partial x} + \frac{\partial \phi}{\partial y}\frac{\partial \eta}{\partial y}; \quad \frac{\partial \phi}{\partial t} + g\eta + \tfrac{1}{2}(\nabla \phi)^2 = G(t) \quad \text{on } z = \eta(x, y, t).$$

An exact solution may be verified to be

$$\phi = -\alpha L \omega z \sin \omega t, \quad \eta = h + \alpha L \cos \omega t;$$

that is the fluid moves a solid body or slug, and $G(t)$ is a complicated function which can be evaluated.

Consider now a small perturbation

$$\phi = -\alpha L \omega z \sin \omega t + \psi, \quad \eta = h + \alpha L \cos \omega t + \zeta.$$

The boundary conditions on ψ are

$$\frac{\partial \psi}{\partial n} = 0 \text{ on the cross-section boundary,}$$

$$\frac{\partial \psi}{\partial z} = 0 \quad \text{on } z = \alpha L \cos \omega t,$$

$$\frac{\partial \psi}{\partial t} + g\zeta - \alpha L\omega \frac{\partial \psi}{\partial z} \sin \omega t + \zeta(-\alpha L\omega^2 \cos \omega t) = 0$$

$$\text{on } z = h + \alpha L \cos \omega t. \quad (3.13)$$

The fourth term in eqn (3.13) arises because the condition is evaluated on the unperturbed boundary, thus introducing a correction $\zeta(\partial^2 \phi/\partial z \, \partial t)$. We also have

$$\frac{\partial \psi}{\partial z} = \frac{\partial \zeta}{\partial t}. \quad (3.14)$$

If we now look for a solution $\psi = f(x, y)H(z, t)$, then separating the variables in $\nabla^2 \psi = 0$ we obtain

$$f_{xx} + f_{yy} + k^2 f = 0, \quad \frac{\partial f}{\partial n} = 0 \text{ on side walls,}$$

$$H_{zz} = k^2 H, \quad H_z = 0 \quad \text{on} \quad z = \alpha L \cos \omega t,$$

together with the free-surface conditions of eqns (3.13) and (3.14) on $z = h + \alpha L \cos \omega t$. This is a homogeneous problem for f and Helmholz equation has an infinite set of discrete positive eigenvalues k^2 for the simple closed domain which is the cross-section of the cylinder. Thus there are non-trivial solutions f satisfying these conditions. In the case of a circular domain of radius a, possible solutions are

$$f = J_m(kr)(A_m \cos m\theta + B_m \sin m\theta),$$

where $J_m'(ka) = 0$ and m is any integer; with $m = 0$ the disturbances are radial.

Solving for H for any cross-section

$$H(z, t) = F'(t)\cosh k(z - \alpha L \cos \omega t),$$

where $F'(t)$ is the derivative of an arbitrary function of t. Substituting in (3.14), and integrating with respect to t,

$$\zeta = f(x, y)F(t)k \sinh kh. \quad (3.15)$$

Substituting this result in (3.13), we obtain

$$\frac{d^2 F}{dt^2} + k \tanh kh(g - \alpha L\omega^2 \cos \omega t)F = 0. \quad (3.16)$$

This is a Mathieu equation and, with $\omega t = \tau$, $\Omega_k^2 = gk \tanh kh$, and

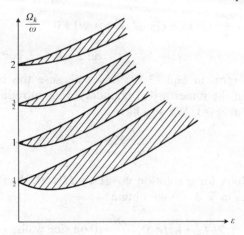

Fig. 3.3. Parameter space for Mathieu equation

$\varepsilon = \alpha L \Omega_k^2 / g$, reduces to the standard form

$$\frac{\mathrm{d}^2 F}{\mathrm{d}t^2} + \left(\frac{\Omega_k^2}{\omega^2} - \varepsilon \cos \tau\right) F = 0. \tag{3.17}$$

The properties of this linear ordinary differential equation are discussed in, for example, Jordan and Smith (1977) and solutions grow in time for certain ranges of the parameters Ω_k/ω and ε, as shown by shaded regions in Fig. 3.3.

We interpret the growth in time of these small perturbations of the free boundary as an instability which will in practice prevent the original quasi-steady state from occurring. For large ε, that is large amplitude oscillations of the cylinder, the free boundary is unstable for almost all values of Ω_k/ω; that is, for almost all wave numbers k. For small ε, that is small amplitude oscillations of the cylinder, the free boundary will be stable except to a disturbance whose natural frequency Ω_k is roughly a multiple of half the frequency ω of the container oscillation. The experiment is easily done and the instability verified, a half-empty beer glass being appropriate apparatus.

An asymptotic analysis of eqn (3.17) for small ε will be briefly described in Chapter 5; a simpler example of the stability of a free boundary in the case of a Hele–Shaw cell is given as Exercise 3.3.

Problem K. Percolation in a sand dune

From a variety of sources, some industrial, large quantities of unwanted water are piped into the sea. The water is usually warmer than the sea, and for a number of reasons it is important to site the waste pipe such that the warmer water is carried away to mix with the cooler salt water, rather than forming a semi-stagnant warm region. Where the sea is shallow, exposing a long sand beach at low tide, the temperature of the wet sand is controlled by the convection of water in it, whereas the temperature of the dry sand depends on conduction and radiation to or from its surface. Thus an important part of the overall heat balance in such a situation is to understand the way in which a sand-bar dries out as the tide recedes. Typically a long beach will dry out in a roughly two-dimensional way; that is, normal to the direction of the tide the beach will have a smooth, slowly varying profile with exposed sand-bars separating liquid streams draining into the sea.

Hence we consider one face of such a sandbar as in Fig. K1, where z is a vertical axis, x is normal to the tide and water is flowing in the y direction as the water level $z = -Z(t)$ rises and falls with the tide. Water will percolate into the sand, leaving a dry region separated from the wet sand by a free boundary QP defined by $z = f(x, t)$. If p is the fluid pressure in the sand above ambient pressure in the air, then for percolation $\nabla^2 p = 0$ and

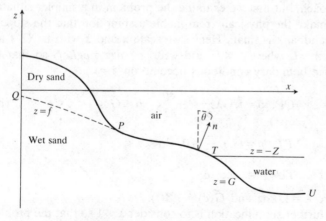

Fig. K1. Cross-section of a sandbar

from eqn (3.12)

$$\text{on } z = f(x, t), \quad p = 0 \quad \text{and} \quad \frac{\partial f}{\partial t} = k\left(\frac{\partial p}{\partial x}\frac{\partial f}{\partial x} - \frac{\partial p}{\partial z} - \rho g\right). \quad \text{(K1)}$$

On the known wetted sand boundary PT (the seepage face) which is above the tide, $z = G(x)$, $p = 0$, and there can only be outflow of water. Hence from Darcy's law

$$\frac{\partial p}{\partial n} \leqslant -\rho g \cos\theta, \quad \text{(K2)}$$

where n is the outward normal and θ its orientation to the vertical. Beneath the water level on the sand boundary TU the pressure is hydrostatic and $p = -\rho g[Z(t) + G(x)]$.

To complete the boundary-value problem we assume a fixed impermeable bottom at $z = -hL$ on which $\partial p/\partial z + \rho g = 0$, together with suitable conditions on $x = 0$, $x = L$ and at $t = 0$. The precise form of these conditions will be established later, but it is clearly necessary for a well-posed problem to close the domain spatially, even if there are no obvious natural boundary conditions. On the free boundary there are two conditions given by eqn (K1), but the end-point P is to be determined. The hope is that the two inequalities, namely $\partial p/\partial n \leqslant -\rho g \cos\theta$ on PT and $f \leqslant G$ on QP, and the boundedness of ∇p, will be sufficient to impose a unique solution and determine the position of P.

The statement of this problem is again suitable for a variational approach, but first we examine the problem in a simpler situation and make the physically reasonable assumption that the slope of the sandbank is small. Hence we scale x and z with L; G, f and Z with εL where $\varepsilon \ll 1$; and write $p + \rho g z = \rho g L \varepsilon \bar{p}$, so that the surface boundary conditions become on $\bar{z} = 0$

$$QP: \quad \bar{p} = \bar{f}, \quad \lambda\frac{\partial \bar{f}}{\partial t} = -\frac{\partial \bar{p}}{\partial \bar{z}}, \quad \bar{p} \leqslant \bar{G}, \quad 0 < \bar{x} < \alpha; \quad \text{(K3)}$$

$$PT: \quad \bar{p} = \bar{G}, \quad \frac{\partial \bar{p}}{\partial \bar{z}} \leqslant 0, \quad \alpha < \bar{x} < c; \quad \text{(K4)}$$

$$TU: \quad \bar{p} = -\bar{Z}, \quad c < \bar{x} < 1,$$

where $\lambda = L/k\rho g$ and $\bar{G}(c) = -\bar{Z}(t)$.

A further simplification is to consider $\lambda \ll 1$ so that the problem is essentially in a steady state for each value of $\bar{Z}(t)$. Then the

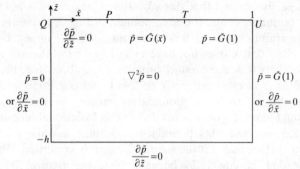

Fig. K2. Boundary value problem for a sandbar

boundary condition on QP is $\partial \bar{p}/\partial \bar{z} = 0$, with the second condition determining $\bar{f}(x)$, and we can choose the origin of $\bar{z}$ so that $\bar{f}(0) = 0$. Initial conditions are now not needed, and two possible conditions on $\bar{x} = 0$ and 1 would seem to model the physical situation reasonably well. The first is the no-flux condition, obtained from the symmetry of a roughly periodic distribution of sandbanks, so that $\partial \bar{p}/\partial \bar{x} = 0$. The second is the hydrostatic pressure condition obtained from considering a single sandbank face abutting on to regions which can provide a small horizontal flow, so that on $x = 0$, $\bar{p} = 0$ but on $x = 1$, $\bar{p} = \bar{G}(1)$. The boundary-value problem is shown in Fig. K2, and it may be that the point T does not exist; that is, the tide has receded to leave no water above the sand in $0 < x < L$.

In this case, if either boundary condition is for zero flux there is a trivial solution $\bar{p} = $ constant with $\alpha = 1$ and no seepage face PT. Thus either the model is inadequate to describe the observed phenomenon, or the boundary-value problem is not unique and it is not a very useful model. With boundary conditions that $\bar{p}$ is prescribed on $\bar{x} = 0$ and 1, the problem is equivalent to the variational inequality

$$\iint \nabla \bar{p} \cdot (\nabla v - \nabla \bar{p}) \, d\bar{x} \, d\bar{z} \geqslant 0, \tag{K5}$$

where the admissible functions v have the properties:

(i) $v = 0$ on $\bar{x} = 0$; $v = \bar{G}(1)$ on $\bar{x} = 1$;
(ii) $v \leqslant \bar{G}$ on $\bar{z} = 0$; and
(iii) they have continuous bounded first derivatives and belong to a convex set.

It is easily verified that this variational inequality is equivalent to the boundary-value problem defined in Fig. K2, together with the constraints $\bar{p} \leq \bar{G}$ in $0 \leq \bar{x} \leq \alpha$ and $\partial \bar{p}/\partial \bar{z} \leq 0$ in $\alpha \leq \bar{x} \leq 1$ on $\bar{z} = 0$. Note that v does not have to satisfy a condition on $\bar{z} = -h$, since $\partial \bar{p}/\partial \bar{z} = 0$ is the *natural* boundary condition for this problem. With $\partial \bar{p}/\partial \bar{x} = 0$ on $\bar{x} = 0$ or $\bar{x} = 1$, no condition would be needed for v on these boundaries either. In Problem J an additional line integral was needed in the variational inequality of eqn (J5) because the boundary condition $\partial \phi/\partial n = \varepsilon$ on the painted surface was not the natural boundary condition. This was not needed in eqn (K5) because it is the natural boundary condition which has to be satisfied on $0 < \bar{x} < \alpha$, $\bar{z} = 0$. In the unsteady problem $(\lambda \neq 0)$ in eqn (K3) a time-stepping procedure may be used, so that at each time step $\partial \bar{p}/\partial \bar{z} = -\lambda (\partial \bar{f}/\partial t)$ is given on $0 < \bar{x} < \alpha(t)$, $\bar{z} = 0$. This will no longer be the natural boundary condition, and eqn (K5) will need to be suitably modified.

For the general problem in which the sandbank slope is not small the use of a Baiocchi transformation, as in Exercise 3.2, leads to a constrained variational principle equivalent to a quasi-variational inequality. This is reviewed in Aitchison, Elliott and Ockendon (1983), where the numerical solution of the variational inequality of eqn (K5) is discussed in detail.

3.4. Application of complex variable methods

In the preceding sections a number of interesting physical problems were modelled by the boundary-value problem of solving Laplace's equation in a rectangular domain, with different boundary conditions on different sections of the boundary. A limiting version of these problems which may be of interest is when the domain is the upper half space $y > 0$, the function vanishes as $y \to \infty$, and different linear boundary conditions are to be satisfied on different sections of the boundary $y = 0$ (as in Exercise 3.4a). Complex-variable theory may be applied in two ways to obtain solutions to problems of this kind, either through the use of Cauchy Principal Value integrals and the Plemelj formulae, or by the use of the complex Fourier transform.

Consider the Cauchy integral

$$F(z) = \frac{1}{2\pi} \int_C \frac{f(t)\, dt}{z - t}$$

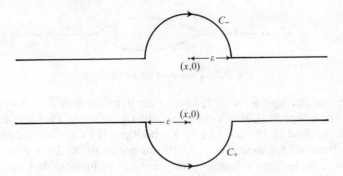

Fig. 3.4. Complex contours for Plemelj formulae

where t and $z = x + iy$ are complex variables, and the contour C is the real axis in the t plane. Assume that f is an analytic function of t, which vanishes as $|t| \to \infty$ on the real axis, so that the integral converges for $z \neq t$ and $F(z) \to 0$ as $|z| \to \infty$. Define $F_+(x)$ and $F_-(x)$ as the limiting values of $F(z)$ as z approaches the real axis from above and below respectively, so that

$$F_\pm(x) = \frac{1}{2\pi} \int_{C_\pm} \frac{f(t)\, dt}{x - t}$$

where C_+ and C_- are the contours shown in Fig. 3.4. Then by Cauchy's residue theorem

$$F_+ - F_- = \frac{1}{2\pi} \int_{C_+ - C_-} \frac{f(t)\, dt}{x - t} = -if(x). \tag{3.18}$$

Also

$$F_+ + F_- = \lim_{\varepsilon \to 0} \left(\frac{1}{\pi} \int_{-\infty}^{x - \varepsilon} \frac{f(t)\, dt}{x - t} + \frac{1}{\pi} \int_{x + \varepsilon}^{\infty} \frac{f(t)\, dt}{x - t} \right),$$

and it is easily shown that this limit exists; it is called the Cauchy Principal Value of the singular integral

$$F_+ + F_- = \frac{1}{\pi} \int \frac{f(t)\, dt}{x - t}. \tag{3.19}$$

Note that the usual definition of an integral would require that the limit is taken omitting the range $(x - \varepsilon, x + \varepsilon')$, where ε and $\varepsilon' \to 0$ independently. This limit does not exist, and the bar on

Fig. 3.5. Flow past an aerofoil

the integral sign is to remind us that it exists only in a special sense. Integrals of this kind have already arisen in Problem F.

Equations (3.18) and (3.19) are the Plemelj formulae, and may be obtained for functions f which are not analytic. They give an almost immediate solution to the problem of inviscid flow past a thin symmetric wing as in Fig. 3.5. If the velocity potential is $Ux + \phi$, then $\nabla^2 \phi = 0$ outside the aerofoil, which if it is thin implies that $\nabla^2 \phi = 0$ everywhere except on the slit $y = 0$, $0 < x < 1$, and $\nabla \phi \rightarrow 0$ far from the slit. The boundary condition of no normal velocity on the upper surface of the wing is $(U + \partial\phi/\partial x) Y'(x) = \partial\phi/\partial y$ on $y = Y(x)$, and for a thin wing this reduces to $\partial\phi/\partial y = UY'(x)$ on $y = 0+$ for $0 < x < 1$. By symmetry $\partial\phi/\partial y = 0$ on the remainder of $y = 0$. The boundary-value problem is therefore to find ϕ satisfying $\nabla^2 \phi = 0$ in the upper half plane, with $\partial\phi/\partial y$ prescribed on $y = 0$ and $\nabla \phi$ vanishing as $x \rightarrow \pm\infty$, $y \rightarrow \infty$.

To solve this problem using Cauchy integrals (and it could be done by Fourier transforms or by using a Green's function as in Exercise 3.1) we observe that ϕ is the real part of some analytic function of z, and hence we may write

$$\frac{\partial\phi}{\partial x} - i\frac{\partial\phi}{\partial y} = F(z) = \frac{1}{2\pi} \int_0^1 \frac{f(t)\,dt}{z - t}.$$

If f is required to be real then $\partial\phi/\partial y = 0$ when $z = x$ for $x < 0$ or $x > 1$, and $\nabla \phi$ vanishes for $|z| \rightarrow \infty$. From eqn (3.18),

$$\left[\frac{\partial\phi}{\partial x} - i\frac{\partial\phi}{\partial y}\right]_{y=0-}^{y=0+} = -if(x), \quad 0 < x < 1, \tag{3.20}$$

so that

$$\left[\frac{\partial\phi}{\partial x}\right]_{y=0-}^{y=0+} = 0 \quad \text{and} \quad \left[\frac{\partial\phi}{\partial y}\right]_{y=0-}^{y=0+} = f(x).$$

Hence choose $f(x) = 2UY'(x)$ so that, using the symmetry, the boundary condition on $y = 0$, $0 < x < 1$ is satisfied. The solution to the boundary-value problem is therefore

$$\phi = \text{Real part of } \left(\frac{U}{\pi} \int_0^1 Y'(t) \log(z - t) \, dt \right)$$

$$= \frac{U}{2\pi} \int_0^1 Y'(t) \log[(x - t)^2 + y^2] \, dt. \tag{3.21}$$

A more interesting application of Cauchy integrals arises from the flow past a wing of zero thickness, so that $\partial\phi/\partial y = UY'(x)$ as $y \to 0\pm$ and

$$\left[\frac{\partial\phi}{\partial y} \right]_{y=0-}^{y=0+} = 0, \quad 0 < x < 1.$$

From eqn (3.20) it is clear that we need to replace $f(t)$ by $ig(t)$ where g is real, so that

$$\frac{\partial\phi}{\partial x} - i\frac{\partial\phi}{\partial y} = F(z) = \frac{i}{2\pi} \int_0^1 \frac{g(t) \, dt}{z - t} \tag{3.22}$$

and $\partial\phi/\partial x = 0$ when $z = x$ for $x < 0$ or $x > 1$. Thus ϕ is constant on $y = 0$ except for $0 \le x \le 1$, and the problem is anti-symmetric. The corresponding boundary-value problem for the upper half space has a boundary condition on $y = 0$ which is in part ϕ prescribed and in part $\partial\phi/\partial y$ prescribed, and as in Problem K the solution will have singularities where the boundary conditions change. Without some additional condition on these singularities the solution will not be unique. For the thin-wing problem this is provided by the Kutta condition for the trailing edge, namely that $\nabla\phi$ is bounded at $x = 1$, together with the requirement that $\nabla\phi$ has an integrable singularity at the leading edge $x = 0$.

From eqn (3.19),

$$F_+ + F_- = \frac{i}{\pi} \int_0^1 \frac{g(t) \, dt}{x - t}, \quad 0 < x < 1,$$

and from eqn (3.22), since g is real,

$$-2UY'(x) = \frac{1}{\pi} \int_0^1 \frac{g(t) \, dt}{x - t}. \tag{3.23}$$

This is a singular integral equation for g, solutions for which are obtained by first considering the homogeneous problem of finding $G(z)$, analytic everywhere except on $y = 0$, $0 < x < 1$, where

$G_+ + G_- = 0$. For details of the theory see Gakhov (1966), but possible functions are clearly

$$G = z^{n_1 + \frac{1}{2}}(z - 1)^{n_2 + \frac{1}{2}}, \tag{3.24}$$

where n_1 and n_2 are any integers, so that there are branch points at $z = 0$ and 1 and a branch cut is taken along $y = 0$, $0 < x < 1$. For any of these functions G we may define $H = GF$ so that

$$H_+ - H_- = G_+(F_+ + F_-) = -2Ui Y'(x)G_+(x)$$

on using eqn (3.23), and from eqn (3.18)

$$H_+ + H_- = G_+(F_+ - F_-) = g(x)G_+(x).$$

Now H also satisfies the Plemelj formulae of eqns (3.18) and (3.19), so that

$$g(x)G_+(x) = \frac{2U}{\pi} \int_0^1 \frac{Y'(t)G_+(t)\,dt}{x - t}$$

is a solution of eqn (3.23) for $g(x)$. Also

$$\frac{\partial \phi}{\partial x} - i\frac{\partial \phi}{\partial y} = \frac{H(z)}{G(z)} = \frac{U}{\pi G(z)} \int_0^1 \frac{Y'(t)G_+(t)}{z - t}\,dt. \tag{3.25}$$

From the conditions for large z, and from eqns (3.24) and (3.25), $n_1 + n_2 + 1 \geq 0$. From the Kutta condition at $x = 1$, $n_2 \leq -1$; and from the leading-edge condition $n_1 \leq 0$. Hence $n_1 = 0$, $n_2 = -1$, and apart from a multiplicative constant G is unique. The solution to the boundary-value problem therefore leads to ϕ being the real part of w, where

$$\frac{dw}{dz} = \frac{Ui}{\pi}\left(\frac{z-1}{z}\right)^{\frac{1}{2}} \int_0^1 Y'(t)\left(\frac{t}{1-t}\right)^{\frac{1}{2}} \frac{dt}{z-t}. \tag{3.26}$$

For further details of the application of the Plemelj formulae to the problem of finding a complex function with a prescribed linear relation between its imaginary parts, and hence of solving the corresponding boundary-value problem for Laplace's equation, reference may be made to Gakhov (1966).

An alternative procedure is to use the Fourier transform defined in Chapter 1 (Section 1.4), but extended to be suitable for functions which do not necessarily satisfy $\int_{-\infty}^{\infty} |f|^2 \, dx < M$. We do this by defining

$$f_+(x) = 0, \qquad x < 0 \qquad f_-(x) = e^{-\gamma_- x} g(x), \quad x < 0$$
$$f_+(x) = e^{\gamma_+ x} g(x), \quad x > 0 \qquad f_-(x) = 0, \qquad x > 0$$

where $\int_{-\infty}^{\infty} |g|^2 \, dx < M$ so that g satisfies eqn (1.7). Then with $k + i\gamma_+ = u$,

$$F_+(u) = \int_0^{\infty} e^{ixu} f_+(x) \, dx$$

and

$$f_+(x) = \frac{1}{2\pi} \int_{-\infty + i\gamma_+}^{\infty + i\gamma_+} e^{-iux} F_+(u) \, du. \qquad (3.27)$$

This is the complex Fourier transform of the function f_+ which can now have linear exponential growth as $x \to \infty$, since γ_+ may be chosen to make the integrals converge. The path of integration of the inversion integral of eqn (3.27) is $\mathrm{Im}\, u = \gamma_+$ (where Im is the imaginary part), and the integral may be evaluated for $x > 0$ by attempting to close the contour in the half plane $\mathrm{Im}\, u < \gamma_+$. Since f_+ vanishes identically for $x < 0$, F_+ can have no singularities in $\mathrm{Im}\, u > \gamma_+$, and closing the contour in this upper half plane gives $f_+ = 0$. A similar discussion leads to

$$f_-(x) = \frac{1}{2\pi} \int_{-\infty - i\gamma_-}^{\infty - i\gamma_-} e^{-iux} F_-(u) \, du, \qquad (3.28)$$

where all the singularities of $F_-(u)$ are in $\mathrm{Im}\, u > -\gamma_-$. These two contours Γ_+ and Γ_- are shown in Fig. 3.6, together with typical singularities of F_+ and F_-. If the singularities of F_+ and F_- are distinct, then it is possible to deform Γ_+ (which may not cross a positive singularity) and Γ_- (which may not cross a negative singularity) into a common contour Γ. Then

$$f(x) = \frac{1}{2\pi} \int_{\Gamma} e^{-iux} F(u) \, du \qquad (3.29)$$

is the inversion integral of the complex Fourier transform

$$\int_{-\infty}^{\infty} e^{iux} f(x) \, dx = F_+ + F_- = F(u),$$

where f may now have linear exponential growth as $x \to \pm\infty$.

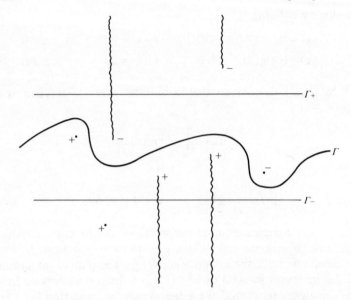

Fig. 3.6. Inversion path for complex Fourier transform

Problem L. Wave reflection from a visco-elastic raft

A proposed method for the protection of oil rigs from rough seas
is to surround them with a thin floating film of visco-elastic
material, which is a relatively cheap by-product of the oil extrac-
tion process. The idea behind this is that the visco-elastic raft will
absorb mechanical energy from the waves and convert it into
heat. There are also proposals for wave-power devices consisting
of a number of hinged rigid rafts which contain springs, whose
motion is used to remove useful energy from the system. The
visco-elastic raft behaves like a large number of hinged rigid rafts
with springs, and has the advantage over other breakwater systems that it is light, easily produced, and does not require large
horizontal forces to tether it.

For simplicity we consider a two-dimensional situation as in
Fig. L1, with water of infinite depth and a raft of width $2L$.

The wave motion is assumed to be modelled by small-
disturbance theory (Stokes waves) as in eqn (1.30), and the raft
modelled as a visco-elastic beam. As in Problem F there will be

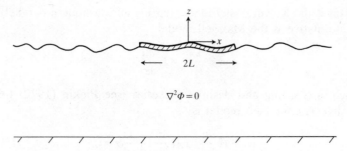

Fig. L1. Visco-elastic raft

two conditions ensuring continuity of pressure and displacement at the raft, on the assumption that there is contact with the water at all points. In suitable non-dimensional variables the boundary-value problem for the velocity potential Φ of water is

$$\nabla^2\Phi = 0, \quad -\infty < z < 0; \quad \frac{\partial\Phi}{\partial z} = 0, \quad z \to -\infty;$$

$$\frac{\partial\Phi}{\partial z} + \frac{\partial^2\Phi}{\partial t^2} = -\frac{\partial P}{\partial t}, \quad \frac{\partial\Phi}{\partial z} = \frac{\partial w}{\partial t}, \quad z = 0. \tag{L1}$$

together with conditions as $z \to \pm\infty$ and at $t = 0$.

For a two-dimensional plate the Bernoulli–Euler model of eqn (1.41) for the non-dimensional displacement $w(x, t)$ of the raft, subject to a pressure force $P(x, t)$, where gravity effects are negligible, is from eqn (F11) given by

$$\frac{\partial^2 w}{\partial t^2} + \bar{\mu}^2 \frac{\partial^4 w}{\partial x^4} = \frac{P}{\gamma}, \tag{L2}$$

where γ is the plate surface density and $\bar{\mu}^2$ is proportional to Young's modulus E. This model for the beam or plate will, however, conserve energy, and has to be modified to represent a visco-elastic raft which dissipates energy. The simplest model of visco-elasticity assumes that stress is a linear function of strain and rate of strain, so that in the one-dimensional case $\sigma = E(\partial u/\partial x) + F(\partial \dot{u}/\partial x)$, which may be rewritten

$$\sigma = E\left(1 + t_1 \frac{\partial}{\partial t}\right) \frac{\partial u}{\partial x},$$

where t_1 is called a relaxation time.

This is the Kelvin–Voigt model (spring and dashpot in parallel); an alternative is the Maxwell model

$$\dot{\sigma} + \frac{1}{t_2}\sigma = E\frac{\partial \dot{u}}{\partial x},$$

which is a spring and dashpot in series (see Pipkin (1972) for details). A combined model is

$$\frac{\partial}{\partial t}\left(\sigma - E\frac{\partial u}{\partial x} - Et_1\frac{\partial \dot{u}}{\partial x}\right) + \frac{1}{t_2}\sigma = 0, \tag{L3}$$

where t_1 and t_2 are relaxation times. With this extended Hooke's law, the relation between the bending moment M and the curvature $\partial^2 w/\partial x^2$ in the derivation of the beam equation in Chapter 1 must be modified appropriately, and a more complicated beam or plate equation derived. However, if we look for solutions periodic in time with frequency ω, then eqn (L3) will become

$$\left(1 + \frac{1}{i\omega t_2}\right)\sigma = E(1 + i\omega t_1)\frac{\partial u}{\partial x},$$

so that we define a complex Young's modulus

$$\tilde{E} = \frac{E(1 + i\omega t_1)}{\left(1 + \dfrac{1}{i\omega t_2}\right)}.$$

The modified form of eqn (L2) becomes

$$-\gamma\omega^2\eta + \beta\frac{d^4\eta}{dx^4} = (p)_{z=0},$$

where $w = \eta(x)e^{i\omega t}$, $P = pe^{i\omega t}$, $\Phi = \phi e^{i\omega t}$ and $\beta = \gamma\bar{\mu}^2(1 + i\omega t_1)/1 + (1/i\omega t_2)]$. The boundary-value problem for ϕ is

$$\nabla^2\phi = 0, \quad -\infty < z < 0; \quad \frac{\partial\phi}{\partial z} = 0, \quad z \to -\infty;$$

$$\frac{\partial\phi}{\partial z} - \omega^2\phi = 0, \qquad z = 0, \quad |x| > 1;$$

$$\frac{\partial\phi}{\partial z} - \omega^2\phi = -i\omega p, \quad z = 0, \quad |x| < 1,$$

where $i\omega\eta = (\partial\phi/\partial z)_{z=0}$.

Eliminating η and p, this condition reduces to

$$\frac{\partial\phi}{\partial z} - \omega^2\phi = 0, \quad z = 0, \quad |x| > 1,$$

$$\frac{\partial\phi}{\partial z} - \omega^2\phi = \gamma\omega^2\frac{\partial\phi}{\partial z} - \beta\frac{\partial^5\phi}{\partial x^4 \partial z}, \quad z = 0, \quad |x| < 1. \tag{L4}$$

To complete the problem we need boundary conditions as $x \to \pm\infty$. Consider the problem of a single wave train $e^{i(\omega t - kx)}$ propagating towards the raft from large negative x, with only an outgoing wave for large positive x. Such a wave can only be propagated if k satisfies the dispersion relation $k = \omega^2$ for Stokes waves on deep water, and we look for a solution

$$\phi = e^{-i\omega^2 x + \omega^2 z} + \psi(x, z).$$

Then

$$\nabla^2\psi = 0, \quad -\infty < z < 0; \quad \frac{\partial\psi}{\partial z} = 0, \quad z \to -\infty;$$

$$x \to +\infty, \text{ transmitted wave only};$$
$$x \to -\infty, \text{ reflected wave only}; \tag{L5}$$

$$\frac{\partial\psi}{\partial z} - \omega^2\psi = m(x), \quad z = 0,$$

where

$$m(x) = \gamma\omega^2\frac{\partial\psi}{\partial z} - \beta\frac{\partial^5\psi}{\partial x^4 \partial z} + (\gamma\omega^4 - \beta\omega^{10})e^{-i\omega^2 x} \text{ for } |x| < 1,$$

$$m(x) = 0 \text{ for } |x| \geq 1. \tag{L6}$$

The boundary-value problem of eqn (L5) is of the type described earlier in this section, where the boundary condition changes its form (but remains linear) on a section of the axis $y = 0$, and a solution is required in a half space. We therefore expect singularities in ψ or its derivatives at $|x| = 1$ and a non-unique solution, as in the problem of flow past a wing of zero thickness. Appropriate conditions are needed at $|x| = 1$ to make the problem well posed, and physically we would expect that both the bending moment and shear on the raft would be zero at its edges;

that is

$$\frac{\partial^2 \eta}{\partial x^2} = 0 = \frac{\partial^3 \eta}{\partial x^3} \quad \text{as} \quad |x| \to 1-,$$

and hence

$$\frac{\partial^3 \psi}{\partial z \, \partial x^2} = \omega^6 e^{-i\omega^2 x}, \quad \frac{\partial^4 \psi}{\partial z \, \partial x^3} = -i\omega^8 e^{-i\omega^2 x} \quad \text{as} \quad |x| \to 1-, \quad z = 0.$$

(L7)

These are further conditions on the boundary-value problem of eqn (L5).

With a boundary condition as complicated as eqn (L6) the use of Cauchy integrals to obtain a solution does not appear to be possible, and in general a numerical solution, either by finite differences or by converting the problem into an integro-differential equation using a Green's function, has to be adopted. However, to give some check on any numerical solution and to check that the problem is well posed we consider the case when γ and $|\beta|$ are small, so that $|A| = |\gamma \omega^4 - \beta \omega^{10}|$ is small and ψ is $O(|A|)$. Then we approximate $m(x)$ by $A \exp(-i\omega^2 x)$, $|x| < 1$, and we have a boundary value problem in which the boundary condition does not change its form but the inhomogeneous term is discontinuous at $|x| = 1$. This is analogous to the problem of the flow past a symmetric thin wing, but because of the form of the boundary condition on ψ it is more straightforward to evaluate it by Fourier transforms than by Cauchy integrals.

If we define $\Psi = \int_{-\infty}^{\infty} e^{iux} \psi(x, z) \, dx$, where u is possibly complex, then $\Psi = B(u) e^{|u|z}$ and B is determined by the condition

$$z = 0, \quad \frac{\partial \Psi}{\partial z} - \omega^2 \Psi = A \left[\frac{e^{-i\omega^2 x + iux}}{iu - i\omega^2} \right]_{-1}^{1} = \frac{2A \sin(u - \omega^2)}{u - \omega^2}.$$

The procedure has introduced the difficulty that $e^{|u|z}$ is not an analytic function of u, and cannot be made analytic without excluding the whole of the imaginary axis so that no inversion path exists. To avoid this difficulty we could either consider a fluid of finite depth h and take the limit as $h \to \infty$ (see Exercise 3.4b), or put $\nabla^2 \psi = \varepsilon^2 \psi$ and let $\varepsilon \to 0$. The latter course is the simpler, and we obtain

$$\Psi = B(u) \exp(u^2 + \varepsilon^2)^{\frac{1}{2}} z.$$

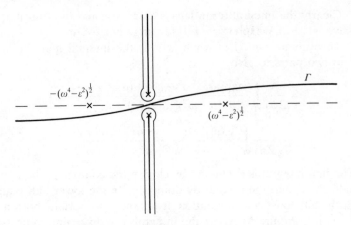

Fig. L2. Complex contour for inversion

This is a well-defined analytic function if we exclude the two branch points $u = \pm i\varepsilon$, and take branch cuts along the imaginary axis above and below the branch points as in Fig. L2.

The inversion contour can now pass from $Re\, u = -\infty$ to $Re\, u = +\infty$ through the small gap between the branch cuts, and the inversion integral is

$$\psi = \frac{A}{\pi} \int_{\Gamma} \frac{\sin(u - \omega^2)\exp[(u^2 + \varepsilon^2)^{\frac{1}{2}}z - iux]}{(u - \omega^2)[(u^2 + \varepsilon^2)^{\frac{1}{2}} - \omega^2]}\, du. \tag{L8}$$

This has poles at $u = \pm(\omega^4 - \varepsilon^2)^{\frac{1}{2}}$, and in the limit $\varepsilon \to 0$ the residues at these poles give contributions proportional to $\exp(\omega^2 z \mp i\omega^2 x)$; that is, the pole at $u = +\omega^2$ represents a transmitted wave, and the pole at $u = -\omega^2$ represents a reflected wave. For $x > 1$ we can close the inversion contour in the lower half plane, with a keyhole contour along the imaginary axis as in Fig. L2, because $e^{\pm iu - iux} \to 0$ for large negative values of Im u. Thus to avoid a reflected wave in $x > 1$ the inversion contour must be below the pole $u = -\omega^2$. A similar argument for $x < -1$, where no transmitted wave is required, implies that Γ is above the pole $u = +\omega^2$ as shown in Fig. L2. This uniquely defines the integral of eqn (L8), and for $x > 1$ we have

$$\psi = -2iA e^{\omega^2(z - ix)} + \frac{Ai}{\pi} \int_0^{\infty} \frac{\sin(\omega^2 + is)}{\omega^2 + is} \left[\frac{e^{isz}}{\omega^2 + isz} - \frac{e^{-isz}}{\omega^2 - isz} \right] e^{-sx}\, ds.$$

Clearly the integral term decays as $x \to \infty$, and the transmitted wave $\sim (1 - 2\mathrm{i}A)\exp[\omega^2(z - \mathrm{i}x)]$ for large positive x.

To evaluate eqn (L8) when $|x| < 1$, the integral must be split into two parts to give

$$\psi = \frac{A}{2\pi \mathrm{i}} \int_\Gamma \frac{\exp[\mathrm{i}(u - \omega^2 - ux) + (u^2 + \varepsilon^2)^{\frac{1}{2}}z]}{(u - \omega^2)[(u^2 + \varepsilon^2)^{\frac{1}{2}} - \omega^2)]}\, \mathrm{d}u$$

$$+ \frac{A}{2\pi \mathrm{i}} \int_\Gamma \frac{\exp[\mathrm{i}(-u + \omega^2 - ux) + (u^2 + \varepsilon^2)^{\frac{1}{2}}z]}{(u - \omega^2)[(u^2 + \varepsilon^2)^{\frac{1}{2}} - \omega^2]}\, \mathrm{d}u.$$

The first integral is evaluated by closing the contour in the upper half plane, and the second by doing this in the lower half plane. Each will have a contribution from the appropriate pole and keyhole contour. At $|x| = 1$ the integrals are divergent, indicating that ψ is singular there, and the form of the singularity can be obtained.

This is not a very valuable solution physically, since there are only small perturbations of the incoming wave train and hence only a small amount of energy of $O(|A|^2)$ is removed by the visco-elastic raft. It does, however, indicate that eqn (L5) is a well-posed boundary-value problem, provided that appropriate singularities are specified at $|x| = 1$, $z = 0$, and that energy is dissipated by the raft.

3.5. Quasi-static continuum models

We have already obtained in eqn (1.25), as a model for the slow two-dimensional flow of a viscous liquid ($Re \to 0$), the biharmonic equation $\nabla^4 \psi = 0$ for the stream function ψ. This is also an elliptic equation, since it can be shown that there are no curves in the x, y plane across which discontinuities in the fourth derivatives of ψ can occur. A general definition of ellipticity, which follows on from the discussion of the hyperbolicity of the system of eqn (2.40), namely

$$A\boldsymbol{u}_x + B\boldsymbol{u}_y = \boldsymbol{d},$$

is to require that there are *no* real roots for λ of $\det(B - \lambda A) = 0$. By writing $\nabla^4 \psi = 0$ as a system of four first-order equations of the above form, it may easily be verified that the function is elliptic.

The biharmonic equation is also satisfied by the Airy stress function A (see Exercise 1.8) for the problem of plane strain in linear-elasticity theory (which is a static situation), or one in which any motion may be neglected in determining the stress field.

Solutions of $\nabla^4 \psi = 0$ may be obtained for simple geometries in much the same way as for $\nabla^2 \psi = 0$, and a well-posed problem, analogous to the Dirichlet problem, is for ψ and $\partial \psi / \partial n$ to be given on a closed boundary. For a free boundary in a slow viscous flow, for example adjacent to a region of constant pressure (say zero), the boundary conditions will be those of zero normal and tangential stress, together with the kinematic condition that fluid particles remain in the boundary. The conditions are complicated, even when reduced to conditions on a fixed boundary for free boundaries which are only small disturbances; and such boundary-value problems usually require extensive numerical effort for their solution.

Many materials are neither purely viscous nor purely elastic, and the constitutive relations between stress and strain for such materials need modification if they are to be appropriately modelled. In Problem L the visco-elastic raft was modelled by a stress which depended on both strain and strain rate; a more general model is to allow the stress to depend on the strain and strain history, so that

$$\sigma_{ij} = \int_{-\infty}^{t} \left[\delta_{ij} \lambda (t - \tau) \dot{\varepsilon}_{kk}(\tau) + 2\mu (t - \tau) \dot{\varepsilon}_{ij}(\tau) \right] d\tau - p\delta_{ij}, \quad (3.30)$$

where $\{\varepsilon_{ij}\}$ is the strain tensor and $\{\dot{\varepsilon}_{ij}\}$ is the rate-of-strain tensor. For a purely elastic material $p \equiv 0$ and λ and μ are constants, so that the constitutive law of eqn (1.37) is recovered. For a viscous incompressible fluid div $\mathbf{q} = \dot{\varepsilon}_{kk} = 0$ and $\mu(t) = \mu_0 \delta(t)$, so that the Newtonian law of eqn (1.23) is recovered. In a one-dimensional situation with $\lambda = p = 0$,

$$\sigma = 2 \int_{-\infty}^{t} \mu(t - \tau) \frac{\partial \dot{u}}{\partial x} d\tau$$

and by a suitable choice of μ this can be written in the form of eqn (L3), the extended Hooke's law used to model the visco-elastic plate, as in Exercise 3.5a.

In a great many industrial processes materials are formed by melting them, and typical examples are to be found in the steel, glass and plastics industries. Since a molten material flows more easily the higher its temperature, a useful constitutive law is that of a viscous fluid with a viscosity μ which decreases with increasing temperature. Experimental evidence indicates that for many molten materials a good approximation to this viscosity-temperature law is $\mu = \mu_0 e^{-\beta T}$, and for almost all such processes the Reynolds number is small so that the equations of slow flow are appropriate.

Ignoring body forces, the slow-flow equations are

$$\frac{\partial p}{\partial x_i} = \mu_0 \frac{\partial}{\partial x_j} \left[e^{-\beta T} \left(\frac{\partial q_i}{\partial x_j} + \frac{\partial q_j}{\partial x_i} \right) \right], \qquad (3.31)$$

and have to be solved, together with the energy equation (1.3) for a viscous fluid and the continuity equation (1.1), for p, T and $\mathbf{q}$. In a two-dimensional situation when a stream function ψ exists, eqn (3.31) reduces to

$$\begin{aligned}
\frac{1}{\mu_0} \frac{\partial p}{\partial x} &= \frac{\partial}{\partial x} \left(e^{-\beta T} \frac{\partial^2 \psi}{\partial x \, \partial y} \right) + \frac{\partial}{\partial y} \left[e^{-\beta T} \left(\frac{\partial^2 \psi}{\partial y^2} - \frac{\partial^2 \psi}{\partial x^2} \right) \right], \\
\frac{1}{\mu_0} \frac{\partial p}{\partial y} &= \frac{\partial}{\partial x} \left[e^{-\beta T} \left(\frac{\partial^2 \psi}{\partial y^2} - \frac{\partial^2 \psi}{\partial x^2} \right) \right] - \frac{\partial}{\partial y} \left(e^{-\beta T} \frac{\partial^2 \psi}{\partial x \, \partial y} \right).
\end{aligned} \qquad (3.32)$$

The energy equation, for Prandtl numbers which are not large, reduces to

$$k \nabla^2 T + \mu_0 e^{-\beta T} \left[\left(\frac{\partial^2 \psi}{\partial x^2} \right)^2 + \left(\frac{\partial^2 \psi}{\partial y^2} \right)^2 + 2 \left(\frac{\partial^2 \psi}{\partial x \, \partial y} \right)^2 \right] = 0. \quad (3.33)$$

This problem for p, T and ψ is still elliptic, and we might expect to obtain a well-posed boundary-value problem if T, ψ and $\partial \psi / \partial n$ are given on a closed boundary. However, the problem is now non-linear and a simple example demonstrates that the boundary-value problem can be ill posed if β is large enough.

Consider the flow in a channel between two parallel walls, distance $2h$ apart, with constant temperature on both plates. We look for a solution which is a function of y alone, so that the walls are necessarily streamlines, and by symmetry need only consider half the channel. Equations (3.32) and (3.33) become

$$\frac{\partial}{\partial y} \left(e^{-\beta T} \frac{\partial^2 \psi}{\partial y^2} \right) = P,$$

where P is the pressure gradient (constant) down the channel, and

$$k\frac{\partial^2 T}{\partial y^2} + \mu_0 e^{-\beta T}\left(\frac{\partial^2 \psi}{\partial y^2}\right) = 0.$$

Simple physical boundary conditions are that on $y = h$, $T = T_0$, $\partial\psi/\partial y = 0$ and on $y = 0$, $\partial T/\partial y = 0 = \psi$, where the symmetry about $y = 0$ has been used. With $\theta = \beta(T - T_0)$, $\psi = e^{\beta T_0}Ph^3\phi$ and $y = hz$, the boundary-value problem becomes

$$\frac{\partial^2 \phi}{\partial z^2} = e^\theta z, \quad \phi(0) = 0 = \frac{\partial\phi}{\partial z}(1); \tag{3.34}$$

$$\frac{\partial^2 \theta}{\partial z^2} + \alpha^2 z^2 e^\theta = 0, \quad \theta(1) = 0 = \frac{\partial\theta}{\partial z}(0), \tag{3.35}$$

where $\alpha = Ph^2(\mu_0\beta e^{\beta T_0}/k)^{\frac{1}{2}}$.

We cannot integrate eqn (3.35) in closed form, but a similar equation and boundary conditions, which could be integrated in closed form, was given as Exercise 1.2b. It had the property that for $\alpha > \alpha_c$, where α_c is a critical value of α which can be determined, there are no solutions to the problem, and for $\alpha < \alpha_c$ there are two possible solutions. A numerical integration of eqn (3.35) confirms that there is a critical value of α beyond which the numerical procedures fail to give a solution, and experimentally there appear to be no steady flows for large enough α. We may represent the possible solution states, as the parameter α varies, by evaluating the temperature at the centre line $\theta(0)$ and plotting it against α as in Fig. 3.7.

This graph has the same qualitative properties as for the problem in Exercise 1.2b, and similar results would occur for the more general equation $\partial^2\theta/\partial z^2 + f(z)e^\theta = 0$, provided f is positive. The second solution for $\alpha < \alpha_c$, which does not tend to zero as α tends to zero, turns out to be unstable and is not realized in practice. The boundary-value problem is clearly ill posed, since it does not have a unique solution except for $\alpha = 0$ or α_c, and for the latter values small changes in α would prove to be highly singular. The physical phenomenon, in which a steady flow in a channel is no longer possible when the channel walls are above a certain critical temperature related to α_c, is called *thermal runaway* and the implication is that the unsteady problem 'blows up', that is the temperature grows unboundedly in a finite time.

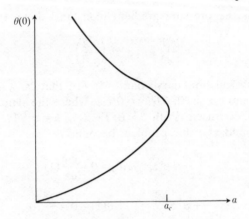

Fig. 3.7. Centre-line temperature distribution

At the end of Problem H a model for two-dimensional plastic flow was proposed in which the material 'slipped' in a direction such that the shear was equal to a constant c, and was less than or equal to c in all other directions. In the notation of Problem H this implied that $\tau = c \sin 2\psi$, and from eqn (H5)

$$|\underline{\underline{\sigma}}|^2 = \tau^2 + \left(\frac{\sigma_1 - \sigma_2}{2}\right)^2 = c^2,$$

where $|\underline{\underline{\sigma}}|$ is called the equivalent stress, and the double underlining indicates that $\underline{\underline{\sigma}}$ is a tensor. In tensor notation

$$|\underline{\underline{\sigma}}|^2 = \sigma_{12}^2 + \left(\frac{\sigma_{11} - \sigma_{22}}{2}\right)^2 = c^2, \tag{3.36}$$

which is invariant under a rotation of coordinate axes (Exercise 3.5b). In Problem H and Exercise 2.7 this compatibility relation led to a hyperbolic model for plastic flow.

For a visco-plastic material the dependence of shear on rate of strain and pressure is still appropriate, but the viscosity may decrease not only with increasing temperature but also with the state of stress. For an isotropic material, the viscosity must be a function of a combination of the stress components which is invariant under a rotation of axes. Thus for a two dimensional visco-plastic situation we define a 'power-law fluid' to be given by $\mu = \mu_0 |\underline{\underline{\sigma}}|^{1-n}$, where $n = 1$ corresponds to a Newtonian viscous fluid. If the stress field is such that μ is not small then, for large enough n, $|\underline{\sigma}|$ is effectively constant and we recover the equation (3.36) for plasticity. This limit is however highly singular, since for finite n the visco-plastic problem is elliptic, and hence in the limit the character of the equations changes.

The visco-plastic equations are conveniently written using both the stream function ψ and Airy stress function A, so that both conservation of mass and momentum are identically satisfied; and we recall that ψ is only defined to within an arbitrary constant, and A to within a linear function of x, y, and a constant. Then, if $\mu = \frac{1}{2}F(|\underline{\sigma}|)$ and $\lambda = 0$ in the constitutive relations of eqn (1.37),

$$2\sigma_{12} = -2\frac{\partial^2 A}{\partial x\,\partial y} = F(|\underline{\sigma}|)\left(\frac{\partial^2\psi}{\partial y^2} - \frac{\partial^2\psi}{\partial x^2}\right), \qquad (3.37)$$

and

$$\sigma_{11} - \sigma_{22} = \frac{\partial^2 A}{\partial y^2} - \frac{\partial^2 A}{\partial x^2} = 2F(|\underline{\sigma}|)\frac{\partial^2\psi}{\partial x\,\partial y},$$

where

$$|\underline{\sigma}|^2 = \left(\frac{\partial^2 A}{\partial x\,\partial y}\right)^2 + \frac{1}{4}\left(\frac{\partial^2 A}{\partial y^2} - \frac{\partial^2 A}{\partial x^2}\right)^2$$

$$= F^2\left[\left(\frac{\partial^2\psi}{\partial x\,\partial y}\right)^2 + \frac{1}{4}\left(\frac{\partial^2\psi}{\partial y^2} - \frac{\partial^2\psi}{\partial x^2}\right)^2\right]. \qquad (3.38)$$

The use of A not only simplifies the use of $|\underline{\sigma}|$ but is also very useful for problems with free boundaries on which there is zero shear and normal stress. Given that $\tau_{nn} = 0 = \tau_{ns}$ on a boundary it can be shown (Exercise 3.5c) that $\partial^2 A/\partial s^2 = 0 = \partial^2 A/\partial n\,\partial s$ and this may be integrated along the boundary to give $A = 0 = \partial A/\partial n$, where two constants of integration have been defined as zero. The third condition that material particles remain on the boundary reduces to $\psi = \text{constant}$.

Problem M. The hot rolling of steel

Steel sheets are formed by squeezing a hot ingot through rollers so that it is flattened into a plate, and then reduced through more rollers into a sheet of the required thickness. Under the rollers the hot metal flows as a visco-plastic material because of the high

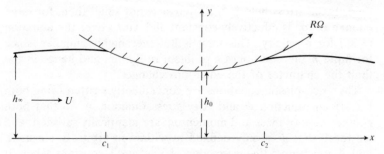

Fig. M1. Metal creep under a roller

stress imposed by the rollers, and it is valuable to be able to predict the deformation, velocity and strain-rate within the metal near the roller. A typical situation under a roller is shown in Fig. M1 with the metal in contact with the roller surface over a section $c_1 < x < c_2$, and a free boundary for $x < c_1$ and $x > c_2$. The incoming strip has speed U and thickness $2h_\infty$; the gap between the pair of rollers (the nip) is $2h_0$, and the rollers have radius R and angular velocities $\pm\Omega$.

For simplicity consider first the case of a viscous fluid $(n = 1)$, which is more relevant to the flow of molten glass and the production of sheets of glass. The thickness of the sheet will be much smaller than the radius of the roller; consider the problem in which the material is only slightly deformed by the roller, that is $\delta = (h_\infty - h_0)/h_\infty \ll 1$. In coordinates made non-dimensional with h_∞ the roller surface is then given by

$$y = 1 - \delta + \frac{x^2 h_\infty}{2R};\qquad\qquad\text{(M1)}$$

we assume that $h_\infty/R = \lambda\delta$, where λ is O(1).

If the stream function Ψ is made non-dimensional with Uh_∞, and the Airy stress function with $\mu_0 Uh_\infty$, then

$$\Psi = y + \delta\psi \quad\text{and}\quad A = \delta a.$$

The boundary conditions on the upper surface of the material may now be reduced to conditions on $y = 1$. For the free boundary sections given by $y = 1 + \delta\eta(x)$, $c_2 < x < c_1$,

$$a = \frac{\partial a}{\partial y} = 0 \quad\text{and}\quad \psi = -\eta.\qquad\qquad\text{(M2)}$$

Under the roller, assuming that there is no slip and that $R\Omega = U(1 + \delta V)$, where V is $O(1)$ (so that the roller-surface and metal-sheet speed are comparable),

$$\psi = 1 - \frac{\lambda x^2}{2}, \quad \frac{\partial \psi}{\partial y} = V. \tag{M3}$$

From the symmetry, the boundary conditions on the centre line $y = 0$ are

$$\psi = 0 = \frac{\partial^2 \psi}{\partial y^2}, \quad \text{implying that} \quad \frac{\partial a}{\partial y} = 0 = \frac{\partial^3 a}{\partial y^3}. \tag{M4}$$

As $x \to -\infty$ there are uniform conditions $\psi \equiv 0 \equiv a$, and we are interested in possible uniform downstream conditions $a \equiv 0$, $\psi = -dy$, $\eta = -d$ as $x \to +\infty$.

At $x = c_1$ and c_2, where the free boundaries attach to the roller, there will be singularities; additional conditions of continuity of stress require that $\partial^2 a / \partial x^2$ and $\partial^2 a / \partial x \, \partial y$ are continuous there. In this linearized problem eqn (M2) then implies that these are both zero.

For the viscous fluid case, and neglecting any variation of viscosity with temperature, both ψ and a satisfy the biharmonic equation; and in addition, from eqn (3.37) with $F = 2\mu_0$,

$$\left(\frac{\partial^2}{\partial y^2} - \frac{\partial^2}{\partial x^2} + 2i \frac{\partial^2}{\partial x \, \partial y} \right) \left(\psi - \frac{ia}{2} \right) = 0. \tag{M5}$$

Boundary-value problems of this kind, involving an infinite strip but with the boundary condition changing its form on a section of the boundary, have been discussed earlier in this chapter but only for Laplace's equation. We cannot expect to be able to find a closed-form solution to this more difficult problem, but Fourier transforms can be used to convert the problem into a pair of dual integral equations; these have a relatively straightforward numerical solution. Defining $\psi^* = \int_{-\infty}^{\infty} \psi(x, y) e^{ikx} \, dx$, and similarly a^*, then

$$\left(\frac{d^2}{dy^2} - k^2 \right)^2 \psi^* = 0$$

and, using the symmetry conditions on $y = 0$, that is with ψ odd in y,

$$\psi^* = By \cosh ky + C \sinh ky.$$

Similarly, but with a even in y,

$$a^* = Dy \sinh ky + E \cosh ky.$$

From eqn (M5),

$$\left(\frac{d}{dy} + k\right)^2 \left(\psi^* - \frac{ia^*}{2}\right) = 0,$$

so that $D = -2iB$ and $E = -2iC$, and the problem is to find the two unknown functions $B(k)$ and $C(k)$. Now for $c_1 < x < c_2$,

$$1 - \frac{\lambda x^2}{2} = \psi(x, 1) = \frac{1}{2\pi} \int_{-\infty}^{\infty} \psi^*(k, 1) e^{-ikx} \, dk,$$

$$V = \frac{\partial \psi}{\partial y}(x, 1) = \frac{1}{2\pi} \int_{-\infty}^{\infty} \frac{\partial \psi^*}{\partial y}(k, 1) e^{-ikx} \, dk,$$

where the Fourier inversion formula has been used and it may be necessary to have an integration path just above the real axis in the complex k plane, to make the integral for ψ^* converge. This may be written in the form

$$\int_{-\infty}^{\infty} M \binom{B}{C} \, dk = \begin{pmatrix} 1 - \dfrac{\lambda x^2}{2} \\ V \end{pmatrix}, \quad c_1 < x < c_2, \qquad \text{(M6)}$$

where

$$M = \frac{e^{-ikx}}{2\pi} \begin{pmatrix} \cosh k & \sinh k \\ \cosh k + k \sinh k & k \cosh k \end{pmatrix}.$$

In addition, on the free boundary

$$\int_{-\infty}^{\infty} N \binom{B}{C} \, dk = \binom{0}{0}, \quad c_1 < x < c_2, \qquad \text{(M7)}$$

where

$$N = \frac{e^{-ikx}}{2\pi} \begin{pmatrix} \sinh k & \cosh k \\ \sinh k + k \cosh k & k \sinh k \end{pmatrix}.$$

Numerical results may be obtained from these dual integral equations, but for a metal the viscous-flow model $n = 1$ is certainly not correct, and for a glass the interesting phenomena occur when the deformation is of the same order as the roller gap. We therefore look for an alternative model based on the assumption that variations in the x direction are much smaller than in the y direction, and δ is not necessarily small. To keep track of the various orders of magnitude we rescale x with $1/\varepsilon$, where $\varepsilon^2 = h_\infty/R \ll 1$, but we do not alter y. Under the roller the boundary conditions become

$$y = h(x) = 1 - \delta + \frac{x^2}{2}, \quad \Psi = 1, \quad \frac{\partial \Psi}{\partial y} = \frac{W}{(1 + \varepsilon^2 h'^2)^{\frac{1}{2}}}, \quad \text{(M8)}$$

where $W = R\Omega/U$ is $O(1)$.

Consider first the case $n = 1$, when equations (3.37) have the form

$$\frac{\partial^2 \Psi}{\partial y^2} - \varepsilon^2 \frac{\partial \Psi}{\partial x^2} = -\varepsilon \frac{\partial^2 A}{\partial x \, \partial y},$$

$$4\varepsilon \frac{\partial^2 \Psi}{\partial x \, \partial y} = \frac{\partial^2 A}{\partial y^2} - \varepsilon^2 \frac{\partial^2 A}{\partial x^2}. \quad \text{(M9)}$$

From the boundary conditions Ψ is $O(1)$, but the size of A is to be determined. Since there are four boundary conditions on Ψ the term $\varepsilon(\partial^2 A/\partial x \, \partial y)$ cannot vanish as $\varepsilon \to 0$, and using both equations this implies that A is $O(1/\varepsilon^3)$. With $a = A\varepsilon^3$, eqns (M9) become

$$\frac{\partial^2 \Psi}{\partial y^2} - \varepsilon^2 \frac{\partial^2 \Psi}{\partial x^2} = -\frac{1}{\varepsilon^2} \frac{\partial^2 a}{\partial x \, \partial y},$$

$$\frac{\partial^2 A}{\partial y^2} - 2 \frac{\partial^2 A}{\partial x^2} = 4\varepsilon^4 \frac{\partial^2 \Psi}{\partial x \, \partial y}. \quad \text{(M10)}$$

We can now construct an asymptotic expansion for Ψ and a in powers of ε, satisfying the conditions of eqn (M4) on $y = 0$. They are

$$a = a_0(x) + \varepsilon^2 \left(a_0''(x) \frac{y^2}{2} + a_1(x) \right) + O(\varepsilon^4),$$

$$\Psi = y b_0(x) - a_0'''(x) \frac{y^3}{6} + O(\varepsilon^2). \quad \text{(M11)}$$

The conditions of eqn (M8) on $y = h(x)$ give

$$\frac{h^2}{3} a_0''' = \frac{1}{h} - W \quad \text{and} \quad 2b_0 = \frac{3}{h} - W.$$

Thus to determine a and Ψ correct to $O(\varepsilon^2)$ we have to obtain a_0 by quadratures, and we need boundary conditions which match the flow under the roller to the flow under the free boundary $y = h(x)$ on which $\Psi = 1$, $A = \partial A/\partial y = 0$. With these boundary conditions the scaling of eqn (M10) is not appropriate, since a_0 is identically zero; and because there are four boundary conditions on A, the term $\varepsilon(\partial^2\Psi/\partial x\,\partial y)$ in eqn (M9) cannot vanish as $\varepsilon \to 0$. We therefore try $\tilde{a} = A/\varepsilon$, so that eqn (M9) becomes

$$\frac{\partial^2\Psi}{\partial y^2} - \varepsilon^2\frac{\partial^2\Psi}{\partial x^2} = -\varepsilon^2\frac{\partial^2\tilde{a}}{\partial x\,\partial y},$$

$$4\frac{\partial^2\Psi}{\partial x\,\partial y} = \frac{\partial^2\tilde{a}}{\partial y^2} - \varepsilon^2\frac{\partial^2\tilde{a}}{\partial x^2}.$$

These two equations have asymptotic solutions

$$\Psi = y\bar{b}_0(x) + O(\varepsilon^2),$$
$$\tilde{a} = \tilde{a}_0(x) + 2\bar{b}_0'y^2 + O(\varepsilon^2).$$

The conditions on $y = h(x)$ require that $\bar{b}_0' = 0 = \tilde{a}_0$ and $h\bar{b}_0 = 1$, so that $\tilde{a} \equiv 0$ and h is constant, correct to the order of any power of ε. Thus the free boundary is $h = 1$ upstream of the roller and is a constant to be determined downstream. The upstream free boundary meets the roller at $x = -(2\delta)^{\frac{1}{2}}$; let the downstream free boundary leave at $x = c$.

At $x = c$ and $-(2\delta)^{\frac{1}{2}}$, we require that the normal stress $\partial^2 A/\partial y^2$ is continuous so that $a_0'' = 0$. Thus c is determined from

$$\int_{-(2\delta)^{\frac{1}{2}}}^{c} \left(\frac{1}{h^3} - \frac{W}{h^2}\right)\mathrm{d}x = 0, \tag{M12}$$

where $h = 1 - \delta + x^2/2$; the strip thickness will decrease if $c < (2\delta)^{\frac{1}{2}}$. The integral can be found in closed form, but gives a complicated expression for c which has to be evaluated numerically. In the case of a small indentation by the roller, when $|\delta - x^2/2| \ll 1$ and $|W - 1| \ll 1$, this integral may be approximated by

$$\int_{-(2\delta)^{\frac{1}{2}}}^{c} \left(-W + 1 + \delta - \frac{x^2}{2}\right)\mathrm{d}x = 0,$$

so that $c/(2\delta)^{\frac{1}{2}}$ is a root of $s^2 - s - 2 + \alpha = 0$, where $\alpha = 3(W-1)/\delta$. Two roots exist if $\alpha < 9/4$, and for $2 < \alpha < 9/4$ both of them are positive. We would expect the metal strip to separate from the roller where $s_0 = x/(2\delta)^{\frac{1}{2}}$ is the smaller of these two roots, giving a strip thickness $h_\infty[1 - \delta(1 - s_0^2)]$. The strip velocity, by conservation of total volume, is $U[1 + \delta(1 - s_0^2)] + O(\delta^2)$.

The asymptotic solution of eqn (M11) is called the *lubrication approximation*, and is usually obtained in u, v, p variables where the pressure p is identified with $-a_0''$; see for example Cameron (1981). However, for the power-law fluid the variables ψ and A are more appropriate and eqn (M10) has the form

$$\frac{\partial^2 \Psi}{\partial y^2} - \varepsilon^2 \frac{\partial^2 \Psi}{\partial x^2} = -|\underline{\underline{\sigma}}|^{n-1} \frac{1}{\varepsilon^2} \frac{\partial^2 a}{\partial x \, \partial y},$$

$$|\underline{\underline{\sigma}}|^{n-1}\left(\frac{\partial^2 a}{\partial y^2} - \varepsilon^2 \frac{\partial^2 a}{\partial x^2}\right) = 4\varepsilon^4 \frac{\partial^2}{\partial x \, \partial y},$$

where

$$|\underline{\underline{\sigma}}|^2 = \frac{1}{\varepsilon^4}\left(\frac{\partial^2 a}{\partial x \, \partial y}\right)^2 + \frac{1}{4\varepsilon^6}\left(\frac{\partial^2 a}{\partial y^2} - \varepsilon^2 \frac{\partial^2 a}{\partial x^2}\right)^2.$$

We proceed with a formal expansion in powers of ε to obtain, as before,

$$a = a_0(x) + \varepsilon^2\left(a_0''(x)\frac{y^2}{2} + a_1(x)\right) + O(\varepsilon^4),$$

where now

$$\Psi = yb_0(x) - \frac{y^{n+2} |a_0'''|^{n-1} a_0'''}{(n+1)(n+2)} + O(\varepsilon^2).$$

Applying the boundary conditions of eqn (M8),

$$|a_0'''|^{n-1} a_0''' = (n+2)\left(\frac{1}{h^{n+2}} - \frac{W}{h^{n+1}}\right).$$

The condition that a_0'' vanishes at $x = -(2\delta)^{\frac{1}{2}}$ and c is

$$\int_{-(2\delta)^{\frac{1}{2}}}^{x^*} \left|\frac{1}{h^{n+2}} - \frac{W}{h^{n+1}}\right|^{1/n} \mathrm{d}x = \int_{x^*}^{c} \left|\frac{1}{h^{n+2}} - \frac{W}{h^{n+1}}\right|^{1/n} \mathrm{d}x,$$

where $Wh(x^*) = 1$. A necessary condition for x^* to exist is $1/W - 1 + \delta > 0$, which reduces to $\alpha < 3/(1-\delta)$. More precise constraints on α may be obtained numerically. This asymptotic procedure may be used in many slow flow problems involving thin layers of material, and will be discussed further in Chapter 5.

Exercises

3.1. (a) Verify that the Green's function for the Dirichlet problem and Laplace's equation on the half space $y > 0$ is

$$-\frac{1}{2\pi}(\log|\boldsymbol{\xi}-\boldsymbol{x}| - \log|\boldsymbol{\xi}+\boldsymbol{x}|).$$

Hence show that the solution to the Dirichlet problem is

$$\phi(x) = \frac{y}{\pi}\int_{-\infty}^{\infty}\frac{h(\xi)\,d\xi}{(x-\xi)^2 + y^2}$$

(b) Show that the Green's function for the Helmholz equation over all *three*-dimensional space is $e^{-kr}/4\pi r$, where $r = |\boldsymbol{x}-\boldsymbol{\xi}|$.

(c) In problem I, consider two spigots on the centre line $y = 1/2$ of a square plate. Verify that $\phi_n^* = 0$ if n is odd, and that only ϕ_0 contributes to E. Hence evaluate E in the form

$$2E = C - \tfrac{1}{3} + \frac{i_1}{2}(1-x_1)^2 + \frac{i_2}{2}(1-x_2^2).$$

If the series for ϕ is truncated after $n = 2$, show that

$$Ri_1 = C - \frac{x_1^2}{2} - 2i_1 G_2(x_1, x_1) - 2i_2 G_2(x_1, x_2),$$

with a similar expression for Ri_2 where G_2 is defined by eqn (I7).

If terms in G_2 may also be neglected show, by using Lagrange multipliers (or otherwise) that E has a minimum where x_1 and x_2 satisfy two simultaneous quartic equations.

3.2. Consider the problem of a porous dam between two reservoirs, with water levels at different heights h_1 and h_2 ($h_1 > h_2$). Assuming that the problem is two-dimensional and that the dam walls are vertical, show that there is a free boundary $z = h(x)$, $0 < x < L$, $h(0) = h_1$, on which $p = 0$ and

$$\frac{\partial p}{\partial z} - h'(x)\frac{\partial p}{\partial x} + \rho g = 0.$$

Show also that on a seepage face $x = L$, $h_2 < z < h(L)$,

$$p = 0 \quad \text{and} \quad \frac{\partial p}{\partial x} \leq 0.$$

Using the Baiocchi transformation

$$u = \int_z^{h(x)} p(x, z') \, dz', \quad z < h; \quad u = 0, \quad z > h;$$

verify that $u = \partial u / \partial z = 0$ on $z = h$ and

$$\frac{\partial^2 u}{\partial x^2} + \frac{\partial^2 u}{\partial z^2} = \alpha, \quad \text{where} \quad \alpha = \rho g, \quad z < h, \quad \alpha = 0, \quad z > h.$$

Show that on $z = 0$, $u = -qx + \frac{1}{2}\rho g h_1^2 = -q(x - L) + \frac{1}{2}\rho g h_2^2$ and find suitable boundary conditions for u on $x = 0$ and $x = L$. Construct a variational inequality equivalent to this boundary value problem for u on the *rectangular* domain $0 < x < L$, $0 < z < h_1$.

3.3. Show that in a Hele–Shaw cell a one-dimensional motion is possible in which there is a free boundary $x = X(t) = \alpha t$ moving with constant speed α, where $p = p_0 + 12\mu\alpha(-x + \alpha t)$ and the fluid lies in $x < \alpha t$. Examine the linear stability of the boundary by writing $x = \alpha t + \varepsilon e^{\sigma t} \sin ny$, n real and positive, and

$$p = p_0 + 12\mu\alpha(-x + \alpha t) + \varepsilon A e^{\sigma t + n(x - \alpha t)} \sin ny.$$

Show that when $\alpha < 0$, so that $\partial p / \partial x > 0$ and the flow region is contracting, the free boundary is unstable for all wave numbers n.

If surface tension is included, so that on the free boundary $p = \delta\kappa$ where κ is its curvature, show that the stability result is unaltered and evaluate A.

For a two-dimensional travelling wave, verify that in coordinates moving with the wave the free-boundary reduces to $\partial p / \partial x \, |\nabla p|^{-2}$ is constant. Show that this may be expressed as $Re(d\zeta/dw) = \text{constant}$, where $\zeta = x + iy$, $w = w(\zeta)$ and $Re \, w = p$.

3.4. (a) In Problem K consider the large width and depth problem, and show that it reduces (in suitable variables) to the boundary-value problem:

$$\nabla^2\phi = 0, \ x, y > 0; \quad \phi = 0, \quad x = 0;$$

$$\frac{\partial \phi}{\partial y} = 0, \quad y = 0, \quad 0 < x < x_0; \quad \phi = G(x), \quad y = 0, \quad x > x_0,$$

where $G(\infty) = 1$, and it has been assumed that $G(x)$ is such that there is only one seepage face $x > x_0$. With $z = x + iy$, $w = \phi + i\psi$, and $\zeta = z^2$, where $\zeta = \xi + i\eta$, show that the boundary conditions on $\eta = 0$ reduce to

$$\phi = 0, \quad \xi < 0; \quad \frac{\partial \phi}{\partial \eta} = 0, \quad 0 < \xi < \xi_0; \quad \phi = \bar{G}(\xi), \quad \xi > \xi_0,$$

and find an appropriate condition on ϕ for large $|\zeta|$.

Make the further transformation $\zeta' = (\zeta - \xi_0)/\zeta$, $F(\zeta') = w(\zeta)/(\zeta')^{\frac{1}{2}}$ and verify that on $\eta' = 0$

$$F_+ = F_- \quad \text{for} \quad \xi' < 0 \quad \text{and} \quad \xi' > 1,$$

$$F_+ - F_- = \frac{2\bar{G}(\xi')}{(\xi')^{\frac{1}{2}}} \quad \text{for} \quad 0 < \xi' < 1.$$

Hence obtain $F(\zeta')$ and define ξ_0 such that $\nabla \phi$ is bounded at $(\xi_0, 0)$.

(b) Consider Problem L for a raft over water of finite depth h, and show that by taking a Fourier transform in the case of small $|A|$

$$\psi = \frac{A}{\pi} \int_C \frac{\sin(u - \omega^2)\cosh[u(z + h)]e^{-iux}}{(u - \omega^2)(u \sinh uh - \omega^2 \cosh uh)} \, du.$$

Verify that there are two poles on the real axis and a pole in every section of the imaginary axis, of length π/h. Describe how to obtain the appropriate inversion contour C for an incident wave in $x < 0$, and obtain the first two terms in a series expansion for ψ in terms of the zeros of $u \tanh uh = \omega^2$.

3.5. (a) Show that, if

$$2\mu(t) = E + Et_1\delta(t) + E(e^{-t/t_2} - 1),$$

then

$$\sigma = 2 \int_{-\infty}^{t} \mu(t-\tau) \frac{\partial^2 u}{\partial x\, \partial t}\, d\tau$$

is equivalent to eqn (L3).

(b) The stress tensor under a rotation of axes L transforms by $\underline{\underline{\sigma}}' = L\underline{\underline{\sigma}}L^{\mathrm{T}}$. Show that the principal stresses are the eigenvalues λ of the matrix $\{\sigma_{ij}\}$. In the two-dimensional case, by considering the quadratic for λ or otherwise, deduce that $\sigma_{11} + \sigma_{22}$ and $\sigma_{11}\sigma_{22} - \sigma_{12}^2$ are invariant. Hence show that $|\underline{\underline{\sigma}}|^2$ defined by eqn (3.36) is invariant and positive definite.

(c) In a two-dimensional situation show that the normal stress on a boundary having orientation θ to the x_1 axis is $\sigma_{11}\sin^2\theta + 2\sigma_{12}\sin\theta\cos\theta + \sigma_{22}\cos^2\theta$, and find the shear. Show that this may be reduced to

$$\frac{\partial^2 A}{\partial s^2} - \frac{\partial \theta}{\partial s}\frac{\partial A}{\partial n},$$

where s is measured along the boundary and n normal to it. Hence verify that the normal and shear stress on a boundary are both zero if $A = \partial A/\partial n = 0$ along it.

(d) A free boundary $y = f(x)$ in a steady two-dimensional viscous flow separates from a fixed rigid wall at the point O. If the x axis is the tangent to the free boundary at O, so that $f(0) = 0 = f'(0)$, look for an asymptotic solution in the form

$$\psi = \psi_0 + \psi_1 + \dots, \quad f = ax^{\lambda+1} + f_1(x) + \dots, \quad (\lambda > 0),$$

valid for small distances r from O.

Show that the boundary conditions on the free streamline may, in polar coordinates, by reduced to

$$\theta = 0, \quad \psi_0 = 0 = \frac{\partial^2 \psi_0}{\partial \theta^2},$$

$$\frac{\partial^3 \psi_0}{\partial \theta^3} + 4\frac{\partial \psi_0}{\partial \theta} - 3r\frac{\partial^2 \psi_0}{\partial r\, \partial \theta} + 3r^2\frac{\partial^3 \psi_0}{\partial r^2\, \partial \theta} = a\delta\lambda(\lambda^2 - 1)r^{1+\lambda},$$

where δ is a suitable surface-tension parameter.

If the fluid is in $y > f$, is bounded by a rigid wall $y = x \tan \alpha$, and has no re-circulating region, show in the case $\delta = 0$ that $\alpha = \pi$ and

$$\psi_0 = Ar^{\frac{3}{2}}\left(\sin \frac{3\theta}{2} + \sin \frac{\theta}{2}\right).$$

From the boundary-value problem for ψ_1 deduce that $\lambda = 1$. If $\delta \neq 0$ show that any separation angle α is possible, provided that λ is a root of

$$\lambda \sin 2\alpha = \sin 2\lambda\alpha.$$

4 Diffusion

4.1. The diffusion equation

In the two preceding chapters we have examined problems which were hyperbolic and elliptic, where for a second-order equation either two or no characteristic directions exist at each point respectively. The third type, for which only one characteristic direction exists, is called *parabolic*, and if this characteristic is given by $t = $ constant then the general linear second-order equation (2.5) may be reduced to

$$\frac{\partial \phi}{\partial t} + q \frac{\partial \phi}{\partial x} = D \frac{\partial^2 \phi}{\partial x^2} + f, \qquad (4.1)$$

where q, D and f are functions of x and t.

This equation models many scientific phenomena, and the variable t is commonly associated with time so that t is increasing and initial data are to be expected on $t = 0$. For many problems the coefficients q and D are constants called the *convective velocity* and *diffusion coefficient*. In general D will be positive, and can thus be scaled to one; also for many problems $q = 0$ so that we obtain the *diffusion equation*

$$\frac{\partial \phi}{\partial t} = \frac{\partial^2 \phi}{\partial x^2} + f. \qquad (4.2)$$

With more space dimensions $\partial^2 \phi/\partial x^2$ is replaced by $\nabla^2 \phi$, and the steady-state diffusion equation reduces to Poisson's equation (discussed in the first section of Chapter 3).

The most common example of diffusion is given by heat conduction in a solid, where conservation of energy as in eqn (1.3) with $q = 0$ leads to

$$\rho c \frac{\partial T}{\partial t} = k \nabla^2 T,$$

often called the *heat conduction equation*. For a moving solid, velocity q, or for heat conduction in an incompressible fluid with negligible viscosity, then from eqn (1.3) a convective term $q \cdot \nabla T$ will be needed, and the convective diffusion equation (4.1) is the appropriate model (with $f = 0$). A further example from fluid mechanics is that of a unidirectional viscous flow $q = [0, 0, w(x, y)]$ such as in a pipe or channel. Then eqn (1.1) is identically satisfied and eqn (1.2) reduces to

$$\rho \frac{\partial w}{\partial t} = \mu \left(\frac{\partial^2 w}{\partial x^2} + \frac{\partial^2 w}{\partial y^2} \right) - \frac{\partial p}{\partial z},$$

where $\partial p / \partial z$ is only a function of time.

It is also easily shown that in a two-dimensional flow the vorticity $\zeta = \partial v / \partial x - \partial u / \partial y$ satisfies a diffusion equation.

A more interesting example is that of molecular diffusion, in which two substances co-exist in a macroscopic element of volume at each point and their relative proportions vary in space and time. In the simplest case one of the substances, called the *matrix*, is fixed and the other diffuses through it with a concentration (amount per unit volume) given by $c(x, t)$. Examples are a dye in a liquid, smoke in the atmosphere, and moisture in a dry solid; further examples are any mixture of substances such as a solute dissolved in a liquid or gas, and a secondary material in a metal forming an alloy. If these mixtures are in equilibrium with c constant, then there is a thermodynamic relationship between the temperature, pressure and chemical potential $A(T, p, c)$ of the mixture, so that the equilibrium concentration will be prescribed in terms of p and T. If there is thermodynamic equilibrium but c varies spatially, then the secondary material or solute will diffuse, and for small concentrations c the diffusional flux is $j = -\alpha \operatorname{grad} A$, where α is positive (so that flux is from high to low potential). Thus $j = -D \operatorname{grad} c$, where $D = \alpha(\partial A / \partial c)$ and is often assumed to be constant. Then for conservation of solute

$$\frac{\partial c}{\partial t} = -\operatorname{div} j = \operatorname{div}(D \operatorname{grad} c), \tag{4.3}$$

which is a diffusion equation. Note that in a liquid with convection of solute $j = cq - D \operatorname{grad} c$, and the convective term $\operatorname{div}(cq)$ appears in eqn (4.3); for an incompressible liquid this reduces to $q \cdot \nabla c$, and c satisfies eqn (4.1) with $f = 0$.

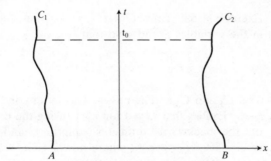

Fig. 4.1. Domain for well-posed parabolic problem

Sufficient conditions for a well-posed boundary-value problem for the diffusion equation (4.2) may be obtained from the maximum-value theorem for the domain shown in Fig. 4.1 with boundaries $t = 0$, $t = t_0(>0)$, C_1, and C_2, where each of the curves C_1 and C_2 intersects lines of constant t only once.

The theorem (for details see for example Protter and Weinberger (1967)) states that for $f \leq 0$ the function ϕ takes its maximum value on $t = 0$, C_1 or C_2. For $f \geq 0$ there is a corresponding minimum-value theorem, and a uniqueness result for the Dirichlet problem follows for $\phi = g(x)$ on $t = 0$, $\phi = h_1(t)$ on C_1 and $\phi = h_2(t)$ on C_2. (If there were two solutions their difference would have both a maximum and a minimum value of zero on the boundary, and hence would be identically zero.) In addition, bounds on small changes in the boundary values for ϕ will be bounds on the consequent changes in ϕ in the interior D. Thus, if solutions to the Dirichlet problem exist with data given on the open boundary, then they are unique and depend continuously on the data, so that the problem is well posed.

For more general boundary data, such as $\partial\phi/\partial n + \alpha\phi$ given on C_1 and C_2, uniqueness may only be established for $\alpha \geq 0$ (n is the outward normal to D). Also note that if we had tried to solve the diffusion equation in $t < 0$, that is backwards in time, with given data at $t = 0$, the maximum-value theorem would allow a maximum to occur on C_1, C_2, or $t = -t_0 < 0$, and no uniqueness result would follow.

For Poisson's equation we defined a Green's function by eqn (3.2) but for the diffusion equation, which is not self-adjoint, it is more straightforward to adopt the definition suggested by eqn

(3.3) or Exercise 1.5a. Thus $G(x, t; \xi, \tau)$ satisfies the adjoint equation in the variables ξ and τ so that

$$\frac{\partial^2 G}{\partial \xi^2} + \frac{\partial G}{\partial \tau} = -\delta(x - \xi)\delta(t - \tau), \tag{4.4}$$

with $G = 0$ on C_1 and C_2 and as $\tau \to \infty$. This latter condition that $G = 0$ as $\tau \to \infty$ implies that $G \equiv 0$ for $\tau > t$ (using the uniqueness theorem on the backwards diffusion equation), and may be replaced by $G = \delta(x - \xi)$ on $\tau = t$ where now $\partial^2 G/\partial \xi^2 + \partial G/\partial \tau = 0$ in $\tau < t$. Applying Green's theorem in the domain bounded by C_1 and C_2, $\tau = 0$ and t,

$$\iint_D Gf \, d\xi \, d\tau = \iint_D \left[G\left(\frac{\partial \phi}{\partial \tau} - \frac{\partial^2 \phi}{\partial \xi^2}\right) + \phi\left(\frac{\partial G}{\partial \tau} + \frac{\partial^2 G}{\partial \xi^2}\right) \right] d\xi \, d\tau$$

$$= \int_{\partial D} \left[\left(G\frac{\partial \phi}{\partial \xi} - \phi\frac{\partial G}{\partial \xi}\right) d\tau + \phi G \, d\xi \right].$$

Evaluating this line integral

$$\phi(x, t) = \iint Gf \, d\xi \, d\tau + \int_A^B gG_{\tau = 0} \, d\xi$$

$$+ \int_{\substack{C_2 \\ 0 < \tau < t}} h_2 \frac{\partial G}{\partial \xi} \, d\tau - \int_{\substack{C_1 \\ 0 < \tau < t}} h_1 \frac{\partial G}{\partial \xi} \, d\tau. \tag{4.5}$$

This gives an explicit expression for ϕ in terms of the boundary data and the Green's function G, which can be shown to exist and to be unique (see for example Stakgold (1979)). It clearly shows that the solution at time t depends only on the data given at earlier times.

G will be singular at $x = \xi$, $t = \tau$ and it is important to obtain the precise form of this singularity, which should be independent of the boundary conditions C_1 and C_2. Thus we can consider the problem in the half space $-\infty < \xi < \infty$, $\tau < t$, and the problem for G is equivalent to solving

$$\frac{\partial \phi}{\partial t} = \frac{\partial^2 \phi}{\partial x^2}, \quad t > 0; \quad \phi = \delta(x), \quad t = 0; \quad \phi \to 0, \quad x \to \pm\infty,$$

if we replace $t - \tau$ by t and $x - \xi$ by x. In a heat-conduction problem this models the temperature distribution in an infinite bar due to a localized 'hot spot' at $x = 0$ and $t = 0$, and is often called the instantaneous heat source temperature distribution. We find ϕ by taking a Fourier transform in x to give

$$\frac{\partial \Phi}{\partial t} = -k^2 \Phi; \quad \Phi = 1, \quad t = 0.$$

Thus

$$\phi = \frac{1}{2\pi} \int_{-\infty}^{\infty} e^{-k^2 t - ikx} \, dx$$

$$= \frac{1}{2\pi} e^{-x^2/4t} \int_{-\infty}^{\infty} \exp\left[-t\left(k + \frac{ix}{2t}\right)^2\right] dk$$

$$= \frac{1}{2\pi t^{\frac{1}{2}}} e^{-x^2/4t} \int_{-\infty + ix/2\sqrt{t}}^{\infty + ix/2\sqrt{t}} e^{-p^2} \, dp,$$

where $p = k + ix/2t$.

By Cauchy's theorem this complex integral is equivalent to the integral along the real axis, because the strip $0 < \mathrm{Im}\, p < x/2\sqrt{t}$ contains no singularities of the integrand, and has the value $\sqrt{\pi}$. Thus the fundamental solution of the diffusion equation is

$$\phi = \frac{e^{-x^2/4t}}{2(\pi t)^{\frac{1}{2}}}, \tag{4.6}$$

which implies that the singular behaviour of G near $x = \xi$, $t = \tau$ is given by

$$G \sim \frac{1}{2[\pi(t-\tau)]^{\frac{1}{2}}} e^{-(x-\xi)^2/4(t-\tau)}, \quad \tau < t,$$

$$G = 0, \quad \tau > t. \tag{4.7}$$

This provides an alternative condition on G to using the δ function, in the same way as with Poisson's equation the condition $G \sim -(2\pi)^{-1} \log |x - x_0|$ was used in Chapter 3.

To construct a Green's function explicitly is not possible except for very simple domains when C_1 and C_2 are lines of constant x representing fixed boundaries. However for such problems, which often occur in applications, taking a Laplace transform in t is a more effective way of constructing explicit solutions. If both C_1 and C_2 are at infinity with $\phi \to 0$ as $x \to \pm\infty$, then either a Laplace transform in t or a Fourier transform in x is appropriate. Given that $\phi = g(x)$ on $t = 0$, then taking a Fourier transform in x

we obtain

$$\Phi^* = G(k)e^{-k^2 t};$$

so that by the convolution theorem and the previous result for the heat source,

$$\phi = \frac{1}{2(\pi t)^{\frac{1}{2}}} \int_{-\infty}^{\infty} g(\xi)e^{-(x-\xi)^2/4t}\,\mathrm{d}\xi. \qquad (4.8)$$

This may be thought of as distributing heat sources at $t = 0$ with density $g(\xi)$; to obtain the result by a Laplace transform see Exercise 4.1a.

With fixed boundaries there are many numerical methods available which use finite differences in both x and t, and they 'march' in time using the property that the solution only depends on what has happened earlier. In this sense parabolic problems are easier to solve by finite difference methods than elliptic problems, where the solution depends on the values at both ends of any finite difference grid.

Problem N. *Smoke dispersion from a high chimney*

Smoke, that is a mixture of gases and particles, diffuses from a tall chimney of height h into the atmosphere and is convected by the prevailing winds. A model is required which will predict the concentration of the smoke at ground level; and of particular interest, for safety reasons, is the maximum value of the concentration of pollutant on the ground. The effects of gravity are unimportant (the Froude number based on the wind speed and chimney height is large), and there is an inversion layer in the lower atmosphere at a height d through which there is no transport of smoke. At the ground the smoke will be deposited at a rate λ which depends on the terrain and the local concentration, and we assume that this rate can be determined empirically. Both the wind velocity and the diffusion coefficient will vary from point to point in a random manner, but we consider length scales such that these quantities take an average value and change only with height. The steady-state convective diffusion equation for the concentration c with diffusion coefficient D is, from eqns (4.1) and (4.3),

$$U\frac{\partial c}{\partial x} = \mathrm{div}(D\,\mathrm{grad}\,c), \qquad (N1)$$

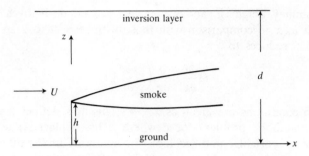

Fig. N1. Smoke dispersion

where U is the mean wind speed which is in the direction x, and z is measured vertically upwards so that U and D are functions of z (see Fig. N1). The concentration c must vanish as $y = \pm\infty$ and we look for the average cross-sectional concentration $\bar{c} = \int_{-\infty}^{\infty} c \, dy$, which by integrating eqn (N1) satisfies

$$U\frac{\partial \bar{c}}{\partial x} = \frac{\partial}{\partial x}\left(D\frac{\partial \bar{c}}{\partial x}\right) - \frac{\partial}{\partial z}\left(D\frac{\partial \bar{c}}{\partial z}\right). \tag{N2}$$

The boundary conditions on $\bar{c}$ are

$$z = 0, \quad \frac{\partial \bar{c}}{\partial z} = \bar{\lambda}; \quad \bar{z} = d, \quad \frac{\partial \bar{c}}{\partial z} = 0; \tag{N3}$$

together with

$$x = 0, \quad \bar{c} = Q\frac{\delta(z-h)}{U(h)}; \quad x \to \infty, \quad \bar{c} \to 0, \tag{N4}$$

where Q is the source strength emanating from the chimney. If we solve this problem for $\bar{c}$ then one way of evaluating c is to assume, because of the random motions of the air, that there is a normal distribution of c in the cross-section; that is,

$$c = \frac{\bar{c}}{\sigma(2\pi)^{\frac{1}{2}}} e^{-y^2/\sigma^2},$$

where σ is to be determined by experiment and may be expected to slowly vary with x and z.

We simplify the two-dimensional problem further by examining the solution on an x length scale which is large compared to the chimney height h, and we neglect second derivatives with respect to x in comparison with those with respect to z so that eqn (N2) reduces to

$$U\frac{\partial \bar{c}}{\partial x} = \frac{\partial}{\partial z}\left(D\frac{\partial \bar{c}}{\partial z}\right), \tag{N5}$$

and no conditions are given as $x \to \infty$. This now defines a well-posed parabolic problem for $\bar{c}$ where x has replaced t as the 'marching' variable, and we look for solutions under different assumptions for d, λ, U and D. The simplest problem is to take $\lambda = 0$, $d \to \infty$ and assume that U and D are independent of z. Then with suitably scaled variables the boundary-value problem is

$$z > 0, \quad x > 0, \quad \frac{\partial \bar{c}}{\partial x} = \frac{\partial^2 \bar{c}}{\partial z^2};$$

$$z = 0, \quad \frac{\partial \bar{c}}{\partial z} = 0; \quad z \to \infty, \quad \frac{\partial \bar{c}}{\partial z} = 0; \tag{N6}$$

$$x = 0, \quad \bar{c} = \delta(z - 1).$$

Define a Laplace transform $\hat{c} = \int_0^\infty e^{-px}\bar{c}(x, z)\,dx$ so that

$$\frac{d^2\hat{c}}{dz^2} - p\hat{c} = -\delta(z - 1)$$

with $d\hat{c}/dz = 0$ on $z = 0$ and $z \to \infty$. The solution is

$$\hat{c} = Ae^{-z\sqrt{p}}, \quad z > 1,$$
$$= B(e^{z\sqrt{p}} + e^{-z\sqrt{p}}), \quad z < 1,$$

where $A = B(e^{2\sqrt{p}} + 1)$ and $[-Ae^{-\sqrt{p}} - B(e^{\sqrt{p}} - e^{-\sqrt{p}})]\sqrt{p} = -1$. Hence

$$\bar{c}_{z=0} = \frac{1}{2\pi i}\int_\Gamma \frac{e^{px - \sqrt{p}}}{\sqrt{p}}\,dp,$$

where Γ is the usual Laplace transform inversion contour. The integrand has a branch point at $p = 0$, and the complex integral may be reduced to an integral along both sides of a branch cut along the negative real axis in the p plane. Thus

$$\bar{c}_{z=0} = \frac{1}{\pi}\int_{-\infty}^{\infty} e^{-s^2x - is}\,ds = \frac{1}{(\pi x)^{\frac{1}{2}}} e^{-1/4x}, \tag{N7}$$

which has a maximum value of $(2/\pi e)^{\frac{1}{2}}$ at $x = \frac{1}{2}$. In dimensional variables the maximum average ground concentration is (Q/U) $(2/\pi e)^{\frac{1}{2}}$ at a distance $Uh^2/2D$ from the chimney, and the assumption that the x scale is larger than the z scale requires that $Uh/D \gg 1$.

The solution obtained by using a Laplace transform in x can also be carried out in the case of finite d, as in Exercise 4.1b. More important is to consider the case when the deposition rate λ is not negligible. For small concentrations the deposition rate will be proportional to the concentration so that $\lambda = \bar{c}v_0(x)$, where $v_0(x)$ is the deposition velocity which varies with the terrain. The problem can no longer be solved by a Laplace transform in x (if $d \to \infty$, a Fourier cosine transform in z may however be used) and a solution is sought in terms of a Green's function.

Thus we look for a function $G(x, z; \xi, \zeta)$ satisfying, as in eqn (4.4), the adjoint equation

$$\frac{\partial^2 G}{\partial \zeta^2} + \frac{\partial G}{\partial \xi} = 0, \quad \xi < x,$$

where $G = \delta(z - \zeta)$, $x = \xi$; $\partial G/\partial \zeta = 0$, $z = 0$ and $z = d$. Then modifying eqn (4.5) for the Neumann boundary condition of eqn (N3),

$$\bar{c}(x, z) = \int_0^x \left(G \frac{\partial \bar{c}}{\partial \zeta} \right)_{\zeta=0} d\xi + \int_0^d (G\bar{c})_{\xi=0} \, d\zeta$$

$$= -\int_0^x G(x, z; \xi, 0)v_0(\xi)\bar{c}(\xi, 0) \, d\xi + G(x, z; 0, 1).$$

Evaluating this on $z = 0$, an integral equation is obtained for the average ground concentration $u(x) = \bar{c}(x, 0)$ in the form

$$u(x) = -\int_0^x G(x, 0; \xi, 0)v_0(\xi)u(\xi) \, d\xi + G(x, 0; 0, 1). \quad (\text{N8})$$

This is a linear Volterra integral equation, which may easily be solved by a numerical iterative process if $v_0(x)$ and G are known.

If $d \to \infty$, G may be found by the method of images in the form

$$G = \frac{1}{2[\pi(x - \xi)]^{\frac{1}{2}}} \{\exp[-(z - \zeta)^2/4(x - \xi)] + \exp[-(z + \zeta)^2/4(x - \xi)]\},$$

and in principle, for finite d, G can be obtained from an infinite series of images whose contributions decrease exponentially. It is a simple matter to check that in the case $v_0 = 0$ the result of eqn (N8) reduces to eqn (N7) obtained by a Laplace transform.

If U and D can be represented by powers of z, then in the case $v_0 = 0$ and $d \to \infty$ a solution may be obtained by a Laplace transform in x, although the inversion integral is difficult to evaluate because the integrand involves Bessel functions of imaginary argument and fractional order (see Exercise 4.1c).

4.2. Stefan problems

Almost all materials can exist in solid, liquid or gaseous form depending on their temperature and pressure. For most pure materials under constant pressure there is a prescribed melting temperature, above which the solid phase becomes liquid, and a vaporization (or boiling) temperature above which the liquid phase becomes a gas. Other less obvious phase changes may also occur in which material properties change discontinuously on a laboratory time-scale. Energy in the form of heat is required to change the phase from solid to liquid and from liquid to gas, and heat is liberated in the reverse process. The amount of energy required (or liberated) is called the latent heat L of the phase change. The simplest and most easily observed phase change is from solid to liquid, and in general this will occur across a moving surface whose position is unknown and is to be determined as part of the solution of the problem for the temperature distribution. Such problems are called Stefan problems, and arise in many applications (see for example Crank (1984) and Chalmers (1967)).

If the phase change between a solid and a liquid occurs on a moving plane given by $x = s(t)$, with the solid in $x < s(t)$ as in Fig. 4.2, then the conditions at the boundary are that the temperature in both the solid and liquid phases is the melting temperature T_m, and an energy balance gives the Stefan condition

$$\left[-k \frac{\partial T}{\partial x} \right]_{x=s-}^{x=s+} = L \frac{ds}{dt}, \quad (L > 0). \tag{4.9}$$

Mass must also be conserved, and if the densities of the two phases are the same, which is a reasonable assumption for the

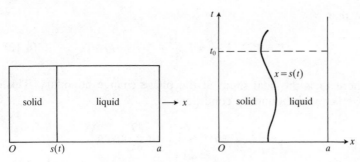

Fig. 4.2. Domain for Stefan problem

solid/liquid interface but not for the liquid/gas interface, then there is no relative motion of the two phases. With the solid at rest and different densities, the fluid is given a velocity $(1 - \rho_s/\rho_l)\, \mathrm{d}s/\mathrm{d}t$ on applying a mass balance. Momentum must also be conserved, and a momentum balance is needed to determine the stress exerted by the fluid on the solid.

With no fluid motion, T satisfies

$$\rho c \frac{\partial T}{\partial t} = k \frac{\partial^2 T}{\partial x^2} \qquad (4.10)$$

in both phases, and for simplicity we shall assume that ρ, c and k are constants with the same values in each phase. Then eqns (4.9) and (4.10), together with the usual boundary conditions for a parabolic problem, complete the statement of the Stefan problem:

$$x = 0, \quad T = f_1(t) < T_m; \quad x = a, \quad T = f_2(t) > T_m; \quad t = 0, \quad T = g(x),$$

where

$$g < T_m \quad \text{for} \quad x < s(0), \quad > T_m \quad \text{for} \quad x > s(0);$$

$$g(0) = f_1(0), \quad g(a) = f_2(a). \qquad (4.11)$$

This is easily interpreted as a sensible physical problem, and the two phases are shown in an x, t diagram in Fig. 4.2.

If the fluid phase is well stirred, or has a turbulent motion so that, by mixing, its temperature is uniform and equal to the melting temperature, then the solid is said to be *ablating*. For a one-dimensional solid block the temperature will satisfy eqn (4.10) together with the boundary conditions on $x = s(t)$, which are that

$$T = T_m, \quad k\frac{\partial T}{\partial x} = L\frac{ds}{dt} + Q(t), \quad Q \geqslant 0, \qquad (4.12)$$

where Q is the heat input at the phase change boundary. This, together with boundary conditions

$$x = 0, \quad T = f(t) < T_m \left(\text{or } \frac{\partial T}{\partial x}\text{ given}\right);$$

$$t = 0, \quad T = g(x) < T_m, \quad \text{where} \quad f(0) = g(0) \quad \text{and} \quad s(0) > 0,$$

is a one-phase Stefan problem. A detailed discussion of the well-posedness of some simple one- and two-phase problems is given by Rubinstein (1971), but for problems of Stefan type there is no general well-posedness result and it is not difficult to construct ill-posed problems.

The simplest exact solution is for a planar phase-change surface moving with constant speed U into a fluid at the melting temperature. Then there is a solution of eqn (4.10) in the form

$$T - T_m = A\left[1 - \exp\left(\frac{\rho c U}{k}(Ut - x)\right)\right], \quad x < Ut, \qquad (4.13)$$

satisfying $T = T_m$ on $x = Ut$.

From the Stefan condition (with no heat input from the fluid), $A = L/\rho c$ and for $U > 0$, $T < T_m$ in the solid. For $U < 0$, should such a situation arise, then $T > T_m$ in the solid phase and the solid is said to be *superheated*.

A two-phase solution with phase-change boundary moving with constant speed has the same form as eqn (4.13), except that A is replaced by A_s in $x < Ut$ and by A_l in $x > Ut$, where $A_s - A_l = L/\rho c$. With $U > 0$ it is now possible to have a situation, e.g. $A_s = 0$, $A_l < 0$, when the fluid is *supercooled*.

For a well-posed problem we require that the phase-change boundary should be stable to small disturbances, and we examine the stability of the one-phase, moving plane-boundary problem by prescribing a small periodic disturbance in the y direction, normal to the motion of the boundary. First the Stefan condition of eqn (4.9) must be obtained for a non-planar boundary defined by $F(\mathbf{x}, t) = 0$. If the normal direction of the surface into the fluid is n, then an energy balance gives

$$\left[-k\frac{\partial T}{\partial n}\right]_s^l \delta t = L\,\delta n,$$

which may be written in the form

$$[k\nabla T]_s^l \cdot \nabla F = L\frac{\partial F}{\partial t}. \tag{4.14}$$

When $F = x - s(t)$ this reduces to eqn (4.9) and, for $F = x - s(t, y)$ and $T \equiv T_m$ in the liquid, reduces to

$$k\frac{\partial T}{\partial x} - k\frac{\partial T}{\partial y}\frac{\partial s}{\partial y} = L\frac{\partial s}{\partial t}. \tag{4.15}$$

Note that this condition is identical in form to that on a free boundary in a Hele–Shaw cell given in eqn (3.11), where $-kT/L$ replaces $\bar{p}/12\mu$.

For a phase-change surface $x = s(y, t) = Ut + \varepsilon e^{\sigma t} \sin ny$, where $\varepsilon \ll 1$ and $n > 0$, we look for a temperature distribution

$$T - T_m = L[1 - e^{U(Ut-x)}] + \varepsilon g(x, t)\sin ny, \quad x < Ut,$$

where for ease of manipulation the variables have been scaled so that $k = \rho c = 1$ in eqn (4.13).

Now T satisfies the two-dimensional heat conduction equation, so that

$$\frac{\partial^2 g}{\partial x^2} - n^2 g = \frac{\partial g}{\partial t},$$

which has solutions of the form $g = a\exp[\sigma t + m(x - Ut)]$, if

$$m^2 - n^2 = \sigma - mU, \tag{4.16}$$

and if $m > -U$ the perturbations will remain small compared to the undisturbed solutions for $(Ut - x)$ large. The condition $T = T_m$ implies that on $x = Ut$,

$$LU(\varepsilon e^{\sigma t} \sin ny) + \varepsilon a e^{\sigma t} \sin ny = O(\varepsilon^2),$$

so that $LU + a = 0$. The Stefan condition of eqn (4.15) implies that on $x = Ut$,

$$-LU^2 + ma - L\sigma = O(\varepsilon).$$

Eliminating a and substituting σ from eqn (4.16), we obtain

$(m+U)^2 = n^2$. With the positive root $m+U = n$ the condition for large $Ut - x$ is satisfied and $\sigma = -Un$.

The plane phase-change surface is therefore stable to small disturbances if $U > 0$ when the solid is not superheated. With $U < 0$ and the solid superheated the boundary is unstable, and the Stefan problem is ill posed. This stability condition, implying that the free boundary is only stable if the solid phase grows into a region of constant temperature and not vice versa, is directly analogous to the stability of the fluid boundary in a Hele–Shaw cell as discussed in Exercise 3.3. The Hele–Shaw problem is, however, easier since $\bar{p}$ satisfies Laplace's equation and not the two-dimensional diffusion equation. The result that the boundary only moves stably in the direction of decreasing pressure is however identical. The stability of the two-phase boundary for the Stefan problem is examined in Exercise 4.2a.

Another exact solution may be obtained in terms of the similarity variable $\eta = x/2\sqrt{t}$, when the heat-conduction equation in one dimension reduces to $T - T_m = F(\eta)$ and F satisfies

$$\frac{d^2 F}{d\eta^2} + 2\frac{dF}{d\eta} = 0. \tag{4.17}$$

For a phase-change boundary $s = 2\alpha\sqrt{t}$, where α is a constant, and the conditions on the free boundary are

$$\eta = \alpha, \quad F = 0, \quad \left[\frac{dF}{d\eta}\right]_s^l = -2\alpha L, \tag{4.18}$$

where we have again taken $\rho c = 1 = k$ by suitable choice of x and t. Solutions of eqn (4.17) subject to the conditions of eqn (4.18) may be used to solve a number of problems; the simplest is a semi-infinite expanse of liquid at the melting temperature, which at $t = 0$ has its temperature at $x = 0$ lowered below the melting temperature so that (say) $T - T_m = -1$. Further boundary conditions are then $F(0) = -1$, $F(\infty) = 0$. The solution is $F \equiv 0$ for $\eta > \alpha$, giving

$$F = -1 + \frac{\operatorname{erf}\eta}{\operatorname{erf}\alpha}, \quad 0 < \eta < \alpha, \tag{4.19}$$

where α is a root of $2L\alpha\exp(\alpha^2)\operatorname{erf}\alpha = 1$ and it can be shown that there is a unique root for α. This is called the Neumann solution, and it is sketched in Fig. 4.3.

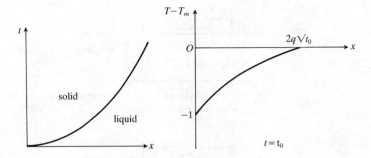

Fig. 4.3. Similarity solution for Stefan problem

A phase may also degenerate (disappear) or be created at a time to be determined. Thus for the ablation problem of eqn (4.12), with boundary condition $\partial T/\partial x = 0$, we may integrate the equation over the domain $0 < t < \tau$, $0 < x < s(t)$ to obtain the line integral

$$\int T \, dx + \int \frac{\partial T}{\partial x} \, dt = 0, \quad (k = \rho c = 1).$$

Hence

$$\int_0^{s(\tau)} (T - T_m) \, dx = \int_0^{s(0)} (g(x) - T_m) \, dx + \int_0^{\tau} q(t) \, dt + [Ls]_0^{\tau}.$$

The phase will degenerate if there exists a value of $\tau > 0$ such that $s(\tau) = 0$, and a necessary condition is that

$$\int_0^{\tau} q(t) \, dt = Ls(0) + \int_0^{s(0)} (T_m - g(x)) \, dx.$$

Clearly for $q > 0$ this will have one root for τ, given that $T_m \geqslant g$.

In our next problem a phase will be created by volumetric heating.

Problem O. The welding of two steel plates

In many manufacturing processes, mild steel plates have to be welded together with great precision and a considerable number of welds may be needed, as for example in making a car body. One convenient way of making an individual weld is to apply two

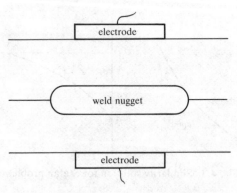

Fig. O1. Welding of two steel plates

electrodes to either side of the plates as in Fig. O1, and to pass a
current. The current will provide a volumetric heat source within
the plates which will melt some of the metal after a sufficient time
has elapsed. Thus a molten nugget will form as in Fig. O1, since
clearly the metal will be cooler at the exposed faces. When the
nugget is large enough to form a sufficiently strong weld, but not
so large that molten metal is in contact with the electrode, the
current is switched off. This welding procedure may be automated and it is important to establish the welding time in terms
of the material properties and the current applied. The problem
is three-dimensional, possibly with axial symmetry about the axis
of the electrodes if they have circular cross-section. However, for
simplicity we shall consider a one-dimensional model which will
give a good approximation to the thickness of the nugget at its
centre if the thickness b of the plates is significantly smaller than
the electrode radius.

The temperature T will satisfy the heat-conduction equation
with a source term $Q = RI^2$, where R is the local resistivity which
may depend on temperature and I is the magnitude of the local
current. However, if an alternating current is used and the
electrodes have a large constant inductance then the current
amplitude is essentially constant. Moreover the time-scale of the
welding process is much longer than the a.c. cycle time, so that
the root mean square value of the current provides the heating
source and may be taken as constant. For mild steel below about
400 °C the resistivity increases with temperature, but above

400 °C (the melting temperature is about 750 °C) it remains roughly constant. Thus there will be a starting problem in which the metal is heated up to 400 °C rapidly because of the growing source term; also current transients and air resistance between the plates, and the electrodes and the plates, will complicate this starting problem. The interesting temperature range is above 400 °C, and we shall model the problem with an initial temperature which is above 400 °C and assume that the time taken to reach this state is small and can be estimated. The thermal conductivities in the molten and solid phases are likely to be very different but for simplicity of the analysis we shall ignore this, although it could be incorporated in a straightforward way.

With non-dimensional variables such that the plates are of thickness 1, the diffusivity is 1, and $u = (T - T_m)/T_m$, then u satisfies

$$\frac{\partial u}{\partial t} = \frac{\partial^2 u}{\partial x^2} + q, \tag{O1}$$

where $q = b^2 RI^2/T_m > 0$. By symmetry $\partial u/\partial x = 0$ at $x = 0$ where the plates touch; and initially at $t = 0$, $u = g(x) < 0$. At $x = 1$, the face of the plate touching an electrode, the boundary condition will be continuity of temperature and heat flow into the electrode. To avoid having to analyse the temperature distribution in the electrode, an empirical heat transfer coefficient H is determined. This gives the boundary condition

$$x = 1, \quad -\frac{\partial u}{\partial x} = H(u - f),$$

where f is some mean temperature of the electrode (<0). To complete the problem a condition on the phase change boundary $u = 0$ is required, and the Stefan condition in these variables is, from eqn (4.9),

$$x = s(t), \quad \left[\frac{\partial u}{\partial x}\right]_{x=s-}^{x=s+} = \lambda \dot{s}, \tag{O2}$$

where $\lambda = L/\rho c T_m > 0$ is the Stefan number. This boundary-value problem is shown in an x, t diagram in Fig. O2, but it is not well posed.

A simple illustration of the lack of well-posedness of this model is to consider the particle which first attains the melting temperature $u = 0$. It will be at $x = 0$, and to change its phase has

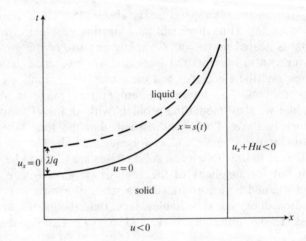

Fig. O2. Domain for welding problem

to acquire an amount of heat λ (in one-dimensional variables) to increase its enthalpy or total heat. Since the particle is at the hottest point it cannot receive heat by conduction, and receives heat only from the source heating. To acquire an amount of heat λ at a rate q will take a time λ/q and its temperature therefore remains constant, equal to zero, for this length of time. Thus a region of constant temperature must exist in the x, t plane of Fig. O2 in which the material is neither totally solid nor molten, called a *mushy* region. The phase change does not therefore take place discontinuously.

An alternative objection to the model with a discontinuous phase change boundary is that it is possible to show (Lacey and Shillor (1983)) that the solid becomes superheated and the phase change boundary is unstable. A mushy region $t_1(x) < t < t_2(x)$, in which $u \equiv 0$, should therefore be introduced into the model where $t_2(0) - t_1(0) = \lambda/q$. If $s_1(t)$ and $s_2(t)$ are the inverse functions of $t_1(x)$ and $t_2(x)$ then, if the solid metal is not to be superheated at any given time,

$$\left[\frac{\partial u}{\partial x}\right]_{s_1-}^{s_1+} \leq 0$$

since $\partial u/\partial x = 0$ in the mushy region. But the enthalpy (total heat) of the material cannot decrease in changing from solid to mush, so that

$$\left[\frac{\partial u}{\partial x}\right]_{s_1+}^{s_1-} \le 0.$$

Hence $[\partial u/\partial x] = 0$ and $\partial u/\partial x = 0$ on $x = s_1(t)$. This argument does not however apply on $x = s_2(t)$, and a discontinuity may occur. From a total heat balance the condition will be

$$x = s_2(t), \quad \frac{\partial u}{\partial x} = -\dot{s}_2[\lambda - q(t_2 - t_1)]. \tag{O3}$$

It may be shown that this is a well-posed problem, and its numerical solution will be discussed in the next section. It is however still a somewhat unsatisfactory model, in that in the mushy region the assumption that $u \equiv 0$ is an approximation to an interesting microstructure in which there is a mixture of solid and liquid phases. This microstructure may be observed; for example 'Tyndall stars' can occur in the melting of ice crystals by sunlight as described in Woodruff (1973), and to describe this phenomenon we shall assume that small regions of liquid grow from a periodic array of nucleation sites. Where melting first occurs, near the *solidus boundary* $x = s_1(t)$, these regions will be small; near the *liquidus boundary* $x = s_2(t)$ there is little solid region left. Thus in the mushy region we have a granular structure as in Fig. O3, although the actual packing of the grains is not known.

If the distance between two nucleation sites is ε and distances are scaled with ε, then there will be small temperature variations within each grain on a scale ε^2 and we write $\bar{u} = u/\varepsilon^2$ so that eqn

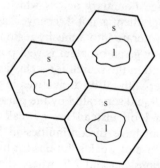

Fig. O3. Granular structure of mushy region

(O1) becomes $\bar{\nabla}^2\bar{u} + q = 0$, where $\bar{\nabla}^2$ is the scaled operator ∇^2. From the Stefan condition of eqn (O2) the outward normal velocity of the solid boundary is given by

$$\lambda\bar{v} = -\left[\frac{\partial\bar{u}}{\partial\bar{n}}\right]_l^s,$$

and on this boundary $\bar{u} = 0$. For a general three-dimensional grain packing this Stefan problem cannot be solved explicitly; but results obtained by averaging over many grains are consistent with the assumptions used earlier in setting up the mushy-region model, and create confidence in it. This microscopic model is examined in a one-dimensional geometry in Exercise 4.2b, and has to be modified near both the solidus and liquidus to ensure that the temperature matches with the temperature in the purely solid and liquid regions. The stability of the phase-change boundary in each grain must also be examined with care, since the model does still predict a small amount of superheating On this microscopic scale, however, surface-energy and surface-tension effects do play a role and it can be shown (see Lacey and Tayler (1983)) that only long-wavelength disturbances can grow, and these have only a finite time in which to grow. This may be an explanation for the star-like structures observed by Tyndall if the grains are essentially two-dimensional.

4.3. Weak solutions.

For Stefan problems the difficulty lies in determining the position of the phase-change boundary across which the temperature is continuous but its gradient is not. Moreover, as in Problem O this boundary may degenerate into a mushy region, and it is not at all obvious whether any given problem is well posed. An alternative approach is to abandon the search for a classical solution which is continuous together with all its derivatives in two regions, that is one solid and one liquid separated by the phase-change boundary or region; we can look instead for a weak solution, that is a function which may have discontinuities in its derivatives. We have discussed the use of weak solutions for hyperbolic equations in Chapter 2, Sections 2.3 and 2.7, where there were difficulties in the non-linear case about the uniqueness of the weak solution when discontinuities (shocks) were present. For some elliptic equations we were able to rephrase boundary value problems as variational principles or variational inequalities in Chapter 3, Section 3.2, but discontinuities in the first derivatives were not possible. However some free-boundary problems, such as Exercise 3.2, could nevertheless be written in a variational, or weak, formulation. For one- or two-phase Stefan problems, not only do weak solutions exist with appropriate discontinuities but they are unique.

We first make the variables non-dimensional as in Problem O, so that the non-dimensional enthalpy is given by

$$h = u, \quad u < 0;$$
$$= u + \lambda, \quad u > 0, \tag{4.20}$$

and for definiteness take boundary conditions from eqn (4.11). We look for a pair of functions (u, h), not necessarily differentiable, satisfying eqn (4.20); then

$$\int_0^{t_0} \int_0^1 \left[h \frac{\partial \phi}{\partial t} + u \frac{\partial^2 \phi}{\partial x^2} \right] dx \, dt = \int_0^{t_0} f_1 \left(\frac{\partial \phi}{\partial x} \right)_{x=0} dt - \int_0^{t_0} f_2 \left(\frac{\partial \phi}{\partial x} \right)_{x=0} dt$$
$$- \int_0^1 h_0(\phi)_{t=0} \, dx \tag{4.21}$$

for all *test functions* ϕ with continuous first derivatives in t and second derivatives in x such that $\phi = 0$ on $x = 0, 1$ and $t = t_0$, and $h_0 = g$ for $g < 0$, $h_0 = \lambda + g$ for $g > 0$. It can be shown (for details see Elliott and Ockendon (1982)) that there is a unique weak solution to this problem which may constructed by the iterative procedure

$$h_{n+1} = h_n + \delta t \left[\frac{\partial^2 u}{\partial x^2} \right]_n, \quad u_n = u(h_n), \tag{4.22}$$

where $[\partial^2 u/\partial x^2]_n$ is some appropriate finite difference representation of $\partial^2 u/\partial x^2$ and the relation $u = u(h)$ is shown in Fig. 4.4.

Let u be a classical solution to this problem, with corresponding enthalpy h; then, using Green's theorem over the domain $0 < t < t_0$, $0 < x < 1$ in Fig. 4.2 and excluding the phase change

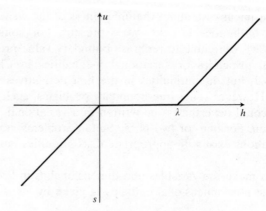

Fig. 4.4. Enthalpy v temperature

boundary $x = s(t)$,

$$\iint \left[\phi\left(\frac{\partial h}{\partial t} - \frac{\partial^2 u}{\partial x^2}\right) + u\frac{\partial^2 \phi}{\partial x^2} + h\frac{\partial \phi}{\partial t} \right] dx\, dt$$

$$= \int \left(u\frac{\partial \phi}{\partial x} - \phi\frac{\partial u}{\partial x} \right) dt - \int \phi h\, dx,$$

where the line integral is taken in an anticlockwise direction around the complete boundary, including both sides of $x = s(t)$. Since $u = 0$ on $x = s(t)$, the contribution from the line integral along both sides of $x = s(t)$ is

$$\int \phi\left(\left[\frac{\partial u}{\partial x}\right]_s^l dt + [h]_s^l\, dx \right),$$

which vanishes if u satisfies the Stefan condition of eqn (4.9); also u satisfies $\partial h/\partial t = \partial^2 u/\partial x^2$, so that evaluating the line integrals on $x = 0$, 1 and $t = t_0$ where $\phi = 0$ we recover eqn (4.21). Hence if u is a classical solution then it is also a weak solution. Because the weak solution is unique any procedure such as that of eqn (4.22) for constructing it must also give the classical solution whenever it exists.

This is the basis of the enthalpy method for the numerical solution of Stefan problems, and the method has the advantage of being on a fixed domain, that is the position of the free boundary emerges as part of the solution and does not modify the iterative procedure of eqn (4.22). Note that a similar advantage arises in using the variational inequality formulation in Chapter 3.

The weak formulation of eqn (4.21) is easily extended to other boundary conditions and to include a source term $q(t)$. The enthalpy method may therefore be applied as in Atthey (1974) to Problem O, for which a stable classical solution does not exist and a region of zero u is obtained as part of the weak solution; it represents the mushy region proposed in the modified classical model for the welding problem. The method may also be applied to problems in more than one space dimension if $[\partial^2 u/\partial x^2]_n$ is replaced by $[\nabla^2 u]_n$.

An important class of generalized Stefan problems arise as models for the phase change of a mixture, such as in the condensation of a gas or the solidification of an alloy. If such a mixture is in thermodynamic equilibrium so that there exists a concentration $c(x, t)$ of the secondary material at each macroscopic point, and the macroscopic time-scale t is much slower than the thermodynamic time-scales, then the 'state' of the element of the mixture is given by its concentration c and temperature T (at normal pressures). Possible states are illustrated by means of a phase diagram which may look like Fig. 4.5a, where the shaded region implies that thermodynamic equilibrium is not possible. There are several qualitative variations to this phase diagram, such as the liquidus curves meeting in a point at the *eutectic temperature* u_e below which any mixture is in a solid phase. It is also possible for the solidus and liquidus to increase with temperature u. However, for small concentrations of secondary material such as in a typical alloy the phase situation may be approximated by the phase diagram shown in Fig. 4.5b, where u is the non-dimensional temperature and the solidus and liquidus curves are straight lines. The crucial feature is that in the shaded region separating two different phases no equilibrium values exist. Thus for the solidification of a binary alloy the region $c > -k_l u$ above the shaded region represents liquid states and for $c < -k_s u$ solid states, and $0 < k_s < k_l$.

If a molten alloy with equilibrium concentration c_l and temperature u is cooled so that the values (c_l, u) lie on the liquidus boundary, then it solidifies to form a solid at this temperature but at a lower concentration c_s. From conservation of mass, secondary material is produced at a rate $c_l - c_s$ at the phase change

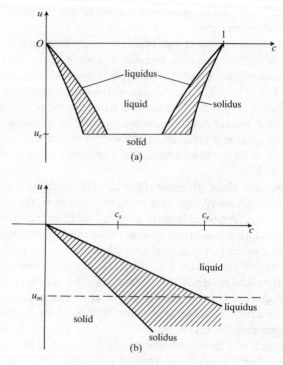

Fig. 4.5. Equilibrium diagrams

boundary, and this must diffuse into the liquid and possibly the solid. The problem is to solve the pair of diffusion equations

$$\frac{\partial c}{\partial t} = D\frac{\partial^2 c}{\partial x^2}, \quad \frac{\partial u}{\partial t} = \frac{\partial^2 u}{\partial x^2} \quad (4.23)$$

in both the solid and liquid phases; it is likely that the non-dimensional diffusion coefficient D in the solid is much smaller than in the liquid, but we have assumed that the liquid and solid densities are the same to avoid liquid motion. The concentrations will, however, differ and a mass balance for the secondary material at the phase change boundary gives

$$x = s(t), \quad \left[D\frac{\partial c}{\partial x}\right]_s^l = -\dot{s}[c]_s^l, \quad (4.24)$$

where the concentrations are determined from the phase diagram

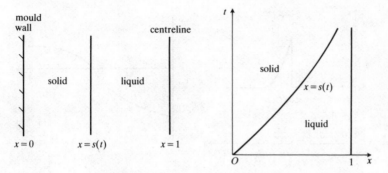

Fig. 4.6. Domain for alloy solidification

by $c_l = -k_l u_m$, $c_s = -k_s u_m$, and $u_m(t)$ is the unknown melting temperature. For definiteness we consider the situation as in Fig. 4.6.

If we assume that $D \equiv 0$ in the solid so that in $x < s(t)$, $c = c(x)$ where $c(s) = -k_s u_m$, then for the liquid at $x = s$

$$D\frac{\partial c}{\partial x} = -\dot{s}(c - c_s), \qquad (4.25)$$

where $c = -k_l u_m = (k_l/k_s)c_s$, and initially $c = c_0$. If $u_m(t)$ were given this would define $c(x)$ in the solid and give a one-phase Stefan problem for c in the liquid region $s < x < 1$. However, u_m is not known and at best this Stefan problem defines a functional relationship between u_m and s; likewise the temperature problem is a two-phase Stefan problem which relates u_m and s. These two functional relations may be sufficient to determine both u_m and s uniquely, but no general results are available. In practice, however, the non-dimensional diffusion coefficient D in the liquid, although much greater than the diffusion coefficient in the solid, is small compared to unity, that is to the non-dimensional heat diffusivity. In this situation an asymptotic solution for c can be obtained, so that with $z = [x - s(t)]/D$,

$$\frac{\partial^2 c}{\partial z^2} = -\dot{s}\frac{\partial c}{\partial z} + O(D), \quad z > 0;$$

where $z = 0$,

$$c = -k_l u_m, \quad \frac{\partial c}{\partial z} = \dot{s}(k_l - k_m)u_m;$$

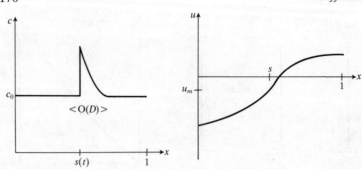

Fig. 4.7. Profiles in alloy solidification

where $z \to \infty$,

$$\frac{\partial c}{\partial z} = 0 \quad \text{or} \quad c \to c_0.$$

This has the solution

$$c = c_0 - e^{-z\dot{s}}(c_0 + k_l u_m), \tag{4.26}$$

and from the condition on $z = 0$, $u_m = -c_0/k_s$ and $c_s = c_0$. The melting temperature is therefore constant to $O(D)$, as is the solid concentration, and the two-phase Stefan problem for the temperature may be solved to find $s(t)$. The form of the solution for c and u is shown in Fig. 4.7.

There is, however, a serious objection to this solution since it necessarily implies that in $x > s(t)$ the concentration rapidly decreases, and hence conditions near to the phase-change boundary in the liquid will lie in the forbidden region of the phase diagram as in Fig. 4.8, where u has been plotted as a function of c from the results shown in Fig. 4.7. In such a situation the liquid is said to be *constitutionally supercooled* and, as in the case of the supercooling of a pure liquid, we may expect that the phase-change boundary is unstable and the solution has been obtained to an ill-posed problem. This instability can be confirmed by a stability analysis similar to that in Exercise 4.2a, and if D is small enough a mushy region forms with no distinct phase-change boundary. Unlike the mushy region in a pure material there can be significant changes in concentration and temperature across the mushy region in an alloy, and to complete the classical

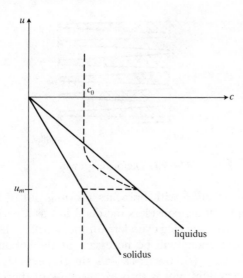

Fig. 4.8. Equilibrium diagram and profile for $D \ll 1$

formulation of the problem a model for the mushy region is required. The non-existence of a phase-change boundary may also be shown by examining the generalized Neumann solution in the case $D \ll 1$ as in Exercise 4.3a.

Similar equations model the condensation of a gaseous binary mixture on a cooled plate, except that for this phase change the liquid density will be much greater than that of the gas. Thus the condensed liquid film creates only a small disturbance to the gas, and the phase-change conditions may be evaluated on the fixed condensation surface. This considerably simplifies the problem, which is discussed further in Exercise 4.3b.

Problem P. Dendritic growth

An alloy ingot may be formed by filling a mould with molten alloy and allowing it to cool by conduction through the walls of the mould. For simplicity we consider a symmetrical one-dimensional geometry as in Fig. 4.6, and from the symmetry $\partial u/\partial x = \partial c/\partial x = 0$ on $x = 1$.

Experiments show that dendrites, that is fingers of solid, grow out from the advancing solid phase, and in general will have an

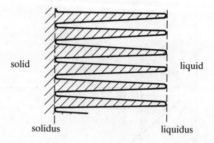

Fig. P1. Dendritic growth

irregular cross-section with branches growing out of each main
stem. A simple model considers the dendrites as a periodic array
of smooth, slowly tapering cylinders for most of their length, as in
Fig. P1; modifications will be necessary at the solidus $x = s_1(t)$
and liquidus $x = s_2(t)$, the root and tip respectively. In some
situations the solidus moves through the bulk of the material, in
others the dendrites are attached to the cooling wall and the
material is nowhere totally solid. The former is the likely situation for the solidification of an alloy, and the conditions at the
root of the dendrite will be eutectic, that is $u = u_e$ in Fig. 4.5a.
For the solidification of water containing a sugar solution, such as
in an ice-lolly, the latter situation is easily demonstrated.

To model this dendritic region we shall, for simplicity, consider
only a two-dimensional situation and assume that there is a
periodic array of dendrites, with axes parallel to the temperature
gradient and a periodic spacing 2ε, where $\varepsilon \ll 1$. This spacing will
be determined by the initial growth of the periodic array from
nucleation sites on the cooled wall, and in this model we shall

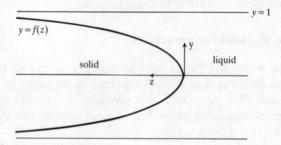

Fig. P2. Cross-section of single dendrite

assume that it is known. Experimental evidence for alloys indicates that $\varepsilon^2 \simeq D$ or $\varepsilon^2 \ll D$, and we would expect the spacing to be related to the wavelengths of unstable linear disturbances on a planar solidification front, which are also $O(D^{\frac{1}{2}})$. We therefore consider the growth of one dendrite as in Fig. P2, and look for a steady shape moving with speed V. With variable y normal to the temperature gradient scaled with ε, and z measured in the negative x direction from the dendrite tip, the governing equations (4.23) become

$$V\frac{\partial u}{\partial z} = \frac{\partial^2 u}{\partial z^2} + \frac{1}{\varepsilon^2}\frac{\partial^2 u}{\partial y^2}, \qquad (P1)$$

$$\delta\frac{\partial c}{\partial z} = \varepsilon^2\frac{\partial^2 c}{\partial z^2} + \frac{\partial^2 c}{\partial y^2}, \qquad (P2)$$

where $\delta = \varepsilon^2 V/D$ and is either $O(1)$ or small and the length scale in the x direction has been chosen so that V is $O(1)$.

The boundary conditions are:

$$y = 1, \quad \frac{\partial c}{\partial y} = \frac{\partial u}{\partial y} = 0; \quad y = 0, \quad \frac{\partial c}{\partial y} = \frac{\partial u}{\partial y} = 0; \qquad (P3)$$

and, if $D \equiv 0$ in the solid dendrite given by $y \leq f(z)$ where $f \equiv 0$ for $z < 0$, then on $y = f(z)$

$$u = u_m(z), \quad c = -k_l u_m(z); \qquad (P4)$$

$$\left[\frac{\partial u}{\partial y} - \varepsilon^2\frac{df}{dz}\frac{\partial u}{\partial z}\right]_s^l = -\varepsilon^2\lambda V\frac{df}{dz}; \qquad (P5)$$

$$\left(\frac{\partial c}{\partial y} - \varepsilon^2\frac{df}{dz}\frac{\partial c}{\partial z}\right)_l = \delta(k_l - k_s)u_m\frac{df}{dz}, \qquad (P6)$$

where eqns (P5) and (P6) are the steady two-dimensional forms of the Stefan conditions of eqns (4.15) and (4.24) and λ is the non-dimensional latent heat.

Finally for simplicity consider a large cooling mould, so that $c = c_\infty$ and $u = u_\infty$ as $z \to -\infty$. There is no obvious physical condition to prescribe behind the dendrite in this travelling-wave solution, although the ellipticity of eqns (P1) and (P2) suggests that conditions are needed, and we first attempt to simplify eqns (P1) to (P6).

We define an average temperature $\bar{u}(z) = \int_0^1 u \, dy$, so that, from eqns (P1) and (P5), $\bar{u}$ satisfies

$$V\frac{d\bar{u}}{dz} = \frac{d^2\bar{u}}{dz^2} + \lambda V\frac{df}{dz} + O(\varepsilon^2). \tag{P7}$$

Moreover, from eqn (P1), $\partial u/\partial y = O(\varepsilon^2)$ so that $u_m(z) = \bar{u}(z)$. Two problems may now be distinguished: the full dendrite $\varepsilon \ll 1$, $\delta \sim 1$; and the 'cell' dendrite where δ is also small and the shape $f(z)$ is slowly varying except near the tip. We examine this latter case in detail and look for a solution to eqn (P2) in powers of $\delta(\gg \varepsilon^2)$ so that

$$c = -k_l\bar{u}(z) + \delta c_1(z, y) + \text{smaller terms.}$$

Substituting this in eqn (P2), and satisfying eqns (P3) and (P4),

$$c_1 = -k_l\frac{d\bar{u}}{dz}\left[\frac{(y-1)^2}{2} - \frac{(f-1)^2}{2}\right],$$

so that from eqn (P6)

$$(k_l - k_s)\bar{u}\frac{df}{dz} = (1-f)k_l\frac{d\bar{c}}{dz}.$$

Integrating, we obtain

$$\bar{u}(1-f)^{1-k_s/k_l} = \text{constant}, \tag{P8}$$

which is called the Scheil equation, and is a statement of conservation of mass. Hence in this dendritic region the heat-conduction equation should be replaced by eqn (P7) where f is determined, except for a constant, from eqn (P8); that is, a non-linear source term is to be included which models the heat release due to the dendritic growth.

The Scheil equation (P8) will not be valid near the tip (or liquidus) where df/dz will become large for small enough z, and we must rescale by defining $z = \delta\bar{z}$ so that, on the scale of $\bar{z}$, $\bar{u} = \bar{u}(0) = u_0$. If $\delta \sim \varepsilon$ all the terms in eqns (P2) and (P6) are of equal order of magnitude, and the only simplification in this tip region is that $u_m(z)$ in eqn (P6) may be replaced by u_0. It is possible to analyse this microstructure in the case $\varepsilon \ll \delta \ll 1$ as in Exercise 4.4b, and to demonstrate that the dendrite tip is parabolic in shape. We shall, however, be content with deriving an expression for the average value of the concentration in the liquid, namely $\bar{c} = \int_f^1 c \, dy$, by integrating eqn (P2) over the domain

$1 > y > f$, $\hat{z} < z < -\infty$, where $\hat{z}$ is O(1). This gives

$$\varepsilon^2 \int_f^1 \frac{\partial c}{\partial z} \, dy = \delta \int_f^1 c \, dy - \delta c_\infty - \delta k_s \int_0^f u_m \, df,$$

where f is evaluated at $z = \hat{z}$. With δ small (and $\delta \sim \varepsilon$) this reduces to

$$-k_l \bar{u}(1-f) = c_\infty + k_s \int_0^f \bar{u} \, df, \tag{P9}$$

which is a mass-conservation statement. Thus if the tip region is considered as a surface of discontinuity at $z = 0$ with a structure in which f changes from zero to f_0 then, taking the limit of eqn (P9) as $z \to 0$, we have $-k_l u_0(1 - f_0) = c_\infty + k_s u_0 f_0$, and

$$f_0 = \frac{c_\infty + k_l u_0}{u_0(k_l - k_s)}. \tag{P10}$$

Finally, from eqn (P7) evaluated across the tip region

$$\left[\frac{d\bar{u}}{dz}\right]_{z=0^-}^{z=0^+} = -\lambda V \frac{(c_\infty + k_l u_0)}{u_0(k_l - k_s)}. \tag{P11}$$

The constant in the Scheil equation (P8) may be determined in terms of u_0 from eqn (P10).

One further condition is, however, needed to determine u_0 and complete the model, but what this condition should be is not clear. Moreover there is a close analogy between the problem and the growth of a single finger-shaped bubble in a Hele–Shaw cell, the so-called Taylor–Saffman problem described by Saffman and Taylor (1958). A similar indeterminacy arises in this simpler problem where the downstream thickness of the bubble as $z \to \infty$ is not determined. There is considerable evidence that this thickness should be half the channel width, but no totally satisfactory arguments have yet been given for this. The dendrite growth problem is harder, and it may be that only a full unsteady model will avoid this indeterminacy. In practice empirical laws are used by metallurgists to complete the steady-state problem.

4.4. Diffusion and convection

The convective-diffusion equation (4.1) arises from heat or mass transfer in a moving matrix, and its properties are less well established than those of the diffusion equation (4.2), If the matrix is a solid, or an incompressible fluid moving in one space dimension only, then the convective velocity q is independent of x. The transformation $x' = x - \int_0^t q(t') \, dt'$, equivalent to taking axes moving with the matrix, reduces eqn (4.1) to the diffusion equation (4.2), and the well-posedness of boundary-value problems for eqn (4.2) can be transferred to eqn (4.1). One example is to move a solid with speed $q(>0)$ past a fixed heat source, which is equivalent to moving the heat source in the opposite direction with speed q through the solid. For a one-dimensional situation with x measured from the heat source, the temperature u satisfies

$$\frac{\partial u}{\partial t} + q \frac{\partial u}{\partial x} = \frac{\partial^2 u}{\partial x^2} + \delta(x), \qquad (4.27)$$

where the diffusivity has been put equal to unity by choice of length-scale. If we consider an infinite solid with $u = 0$ at both ends, there is then no steady-state solution to this problem because heat energy is being supplied at a uniform rate. For the unsteady problem in which the heat source is switched on at time $t = 0$ when the temperature is everywhere zero, the problem is well posed and may be solved by transforms. With q constant we use the Laplace transform $\hat{u} = \int_0^\infty u(x, t) e^{-pt} \, dt$, so that

$$\frac{\partial^2 \hat{u}}{\partial x^2} - q \frac{\partial \hat{u}}{\partial x} - p\hat{u} + \frac{\delta(x)}{p} = 0, \quad \hat{u} \to 0, \quad x \to \pm\infty.$$

This has the solution

$$\hat{u} = A e^{-\alpha x}, \quad x > 0; \quad \hat{u} = A e^{-\beta x}, \quad x < 0,$$

where α and β are the roots of $n^2 + nq - p = 0$ such that $Re \, \alpha > 0$ and $Re \, \beta < 0$, and

$$A = \frac{1}{p(\alpha - \beta)} = \frac{1}{p(q^2 + 4p)^{\frac{1}{2}}}.$$

Hence from the inversion formula for a Laplace transform

$$u = \frac{1}{2\pi i} \int_\Gamma \frac{e^{pt - \alpha x}}{p(q^2 + 4p)^{\frac{1}{2}}} \, dp, \quad x > 0,$$

with a similar expression for $x < 0$ with β replacing α. The integrand has a pole at $p = 0$ and a branch point at $p = -q^2/4$, and by the techniques of contour integration this may be reduced to a real integral along the branch cut and a contribution from the pole. For large time and fixed x, the asymptotic form of u will be given by the contribution from the singularity with largest real part for p, which in this case is the pole. Hence as $t \rightarrow \infty$,

$$u \sim \frac{e^{-qx}}{q}, \quad x<0; \quad u \sim \frac{1}{q}, \quad x>0.$$

The result for $x>0$ is for fixed finite x and does not satisfy the condition $u \rightarrow 0$ as $x \rightarrow \infty$. For a large enough positive value of x this asymptotic form will not be correct, and there will be a region in which the temperature drops from $1/q$ to zero. The asymptotic form can be considered as a steady-state solution if the boundary condition as $x \rightarrow \infty$ is changed to $\partial u/\partial x \rightarrow 0$ from $u \rightarrow 0$. If q depends on time then a Fourier transform in x has to be used, but this leads to less simple integral expressions for u (see Exercise 4.4a).

Examples of the convective-diffusion equation often occur with no source term and convection in one dimension, but diffusion in two space dimensions, so that eqn (4.1) becomes

$$\frac{\partial u}{\partial t} + q \frac{\partial u}{\partial x} = \frac{\partial^2 u}{\partial x^2} + \frac{\partial^2 u}{\partial y^2}. \tag{4.28}$$

The steady-state problem is then elliptic, and is one of the simple elliptic examples already discussed in Chapter 3. Consider the flow of an inviscid fluid past a heated semi-infinite flat plate of constant temperature 1, placed parallel to the stream so that the fluid velocity is everywhere a constant q. Boundary conditions are then that $u = 1$ on $y = 0$, $x>0$ with u vanishing far from the plate. This surely implies that $u \rightarrow 0$ as $x \rightarrow -\infty$ or $y \rightarrow \pm\infty$, but an appropriate condition as $x \rightarrow +\infty$ is not clear as in the previous example. By symmetry $\partial u/\partial y = 0$ on $y = 0$, $x<0$ and it is possible to take a Fourier transform in x to obtain a Wiener–Hopf problem similar to Exercise 2.5b. However, for this simple problem the conformal transformation $\zeta = z^{\frac{1}{2}}$, where $z = x + iy$ and $\zeta = \xi + i\eta$, takes the positive real axis in the z plane into the whole real axis in the ζ plane; this gives boundary conditions $u = 1$ on $\eta = 0$ and $u \rightarrow 0$ as $\eta \rightarrow \infty$. The transformation may be written in the form $x = \xi^2 - \eta^2$, $y = 2\xi\eta$ and the equation becomes (after some simple manipulation)

$$\frac{\partial^2 u}{\partial \xi^2} + \frac{\partial^2 u}{\partial \eta^2} = 2q\left(\xi \frac{\partial u}{\partial \xi} - \eta \frac{\partial u}{\partial \eta}\right).$$

A solution satisfying the boundary conditions may be found in the form $u = u(\eta)$ where

$$\frac{d^2 u}{d\eta^2} + 2q\eta \frac{du}{d\eta} = 0, \quad u(0) = 1, \quad u(\infty) = 1.$$

This may be integrated to give

$$u = 1 - \frac{2}{\sqrt{\pi}} \int_0^{\eta\sqrt{q}} e^{-s^2} \, ds, \tag{4.29}$$

where $(2/\sqrt{\pi}) \int_0^\eta \exp(-s^2) \, ds$ is the error function erf η. Lines of constant temperature are therefore given by $\eta = $ constant, that is the parabolas $y^2 = 4\eta^2(x + \eta^2)$ which are sketched in Fig. 4.9. Note that no boundary condition on u appears to be necessary as $x \to \infty$, despite the ellipticity of eqn (4.25). This situation is analogous to the difficulty of deciding on appropriate boundary conditions as $z \to \infty$ in Problem P, and suggests that no downstream condition on c may be needed in that steady-state problem. Note also that the transformation may be used to solve

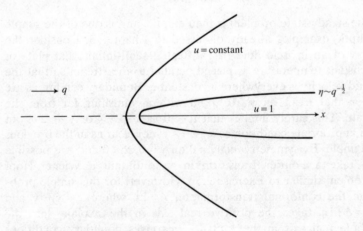

Fig. 4.9. Inviscid flow past a semi-infinite heated flat plate

problems in which the temperature is given on any parabola $\eta = \eta_0$.

If q is large, which implies that the Peclet number $Pe \gg 1$, u will be exponentially small except where $\eta \sim 1/\sqrt{q}$, that is in a parabolic region close to the plate. An alternative approach with q large is to write $x = q\bar{x}$, so that eqn (4.28) reduces to

$$\frac{\partial u}{\partial t} + \frac{\partial u}{\partial \bar{x}} = \frac{\partial^2 u}{\partial y^2} + O\left(\frac{1}{q}\right). \tag{4.30}$$

In the steady state eqn (4.30) is a diffusion equation in which the time-like variable is $\bar{x}$. We would not now expect to be able to pose any boundary condition on u as $\bar{x} \to \infty$, and can assert from its parabolic character that $u \equiv 0$ for $\bar{x} < 0$. Hence appropriate boundary conditions are

$$u = 0, \bar{x} = 0, y > 0; \quad u = 1, y = 0, \bar{x} > 0; \quad u \to 0, y \to \infty, \bar{x} > 0.$$

This has a well-known solution in terms of the similarity variable $\beta = y/2\sqrt{x}$, as in the Neumann solution to a Stefan problem of eqn (4.19). It is

$$u = 1 - \operatorname{erf} \beta, \tag{4.31}$$

which is the appropriate asymptotic expansion of eqn (4.29) for large q.

An alternative derivation of eqn (4.30) was used in Problem N in which q is $O(1)$, but the x length-scale was much greater than the y scale so that $\partial^2 u/\partial x^2$ could be neglected in comparison with $\partial^2 u/\partial y^2$. This is a situation which occurs frequently, and the parabolic version in eqn (4.30) of the elliptic problem of eqn (4.29) is highly advantageous in constructing solutions. Thus in the continuous casting of a metal sheet, the molten metal is drawn from a large well-stirred vat with speed q and is cooled so that it solidifies. With x and y axes as shown in Fig. 4.10 the temperature u satisfies eqn (4.28) in the solidified metal, and we assume that because of the heat mixing $u \equiv 0$ in the liquid. If we look for a steady-state solution there will be a phase-change boundary defined by $y = s(x)$ on which $u = 0$, and the Stefan condition of eqn (4.14) reduces to

$$\lambda q \frac{ds}{dx} = -\frac{\partial u}{\partial y} + \frac{ds}{dx}\frac{\partial u}{\partial x}.$$

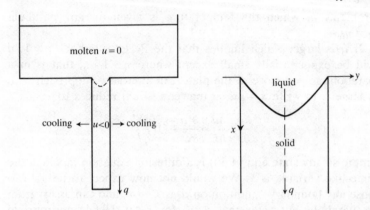

Fig. 4.10. Continuous casting

For a thin metal sheet so that the x scale is much larger than the y scale, or for large Peclet number q, eqn (4.28) reduces to the diffusion equation

$$q\frac{\partial u}{\partial x}=\frac{\partial^2 u}{\partial y^2},$$

with Stefan boundary condition

$$y=s(x),\quad u=0,\quad \frac{\partial u}{\partial y}=-\lambda q\frac{ds}{dx}\bigg/\left(1+\left(\frac{ds}{dx}\right)^2\right).$$

This, together with conditions of cooling on $y=0$ and symmetry on the plate centreline $y=1$, gives a standard one-phase Stefan problem with a stable phase-change boundary as described in Section 4.2, where x is the time-like variable.

Unsteady solutions of eqn (4.30) can be found by taking Laplace transforms in both t and x, but the double inversion integral will be difficult to manipulate. A solution can be obtained by similarity methods for the problem of flow past a semi-infinite heated flat plate in which the heating is switched on time $t=0$. This problem is self-similar in variables $z=y/2\sqrt{t}$, $\tau=x/t$ so that it reduces to

$$\frac{\partial^2 u}{\partial z^2}+2z\frac{\partial u}{\partial z}=4(1-\tau)\frac{\partial u}{\partial \tau},\qquad (4.32)$$

with boundary conditions

$$\tau > 0, z = 0, u = 1; \quad z \to \infty, u \to 0; \quad \tau = 0, u = 0.$$

Two solutions are available. One, independent of τ, is given by $1 - \mathrm{erf}\, z$; the other, independent of t and already obtained in eqn (4.31), is given by $1 - \mathrm{erf}(y/2\sqrt{x}) = 1 - \mathrm{erf}(z/\sqrt{\tau})$. Thus a solution to the problem is

$$
\begin{aligned}
u &= 1 - \mathrm{erf}(z/\sqrt{\tau}), \quad 0 < \tau < 1, \\
u &= 1 - \mathrm{erf}\, z, \quad\quad\ \ \tau > 1,
\end{aligned}
\tag{4.33}
$$

where the two functional forms have the appropriate continuity at $\tau = 1$ because it is a characteristic, so that $\partial u/\partial \tau$ can be discontinuous across it. Equations like (4.32), which are parabolic but have a diffusion coefficient which changes sign so that backward diffusion occurs, clearly need very special boundary conditions in order to be well posed, and few results are available. One such problem is described in Fitt, Ockendon and Shillor (1984).

For a general two-dimensional situation in which the convection velocity has components q_1 and q_2 corresponding to an incompressible fluid matrix, then by continuity $\partial q_1/\partial x + \partial q_2/\partial y = 0$ and in terms of a stream function ψ we have $q_1 = \partial \psi/\partial y$ and $q_2 = -\partial \psi/\partial x$. The convective-diffusion equation may then be written

$$
\frac{\partial u}{\partial t} + \frac{\partial \psi}{\partial y}\frac{\partial u}{\partial x} - \frac{\partial \psi}{\partial x}\frac{\partial u}{\partial y} = \frac{\partial^2 u}{\partial x^2} + \frac{\partial^2 u}{\partial y^2},
\tag{4.34}
$$

where the diffusivity has again been put equal to unity. It is difficult to find closed-form solutions of eqn (4.34) except in a number of limiting cases when the Peclet number, that is the relative size of the convection terms to the diffusion terms, is either large or small.

Problem Q. The shape of laser melt-pools

A technique for making an accurate continuous weld is to use a laser beam as a heat source, applied in a direction normal to the plates to be welded. The laser, which may be considered as a circular cylinder of radius a and prescribed temperature T_1, creates a hot-spot through the plates surrounded by a region of

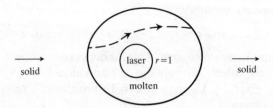

Fig. Q1. Laser melt-pool

molten metal. As the laser moves the shape of this molten region, called a melt-pool, determines the strength of the weld. We consider only the simplest problem in which the laser moves with constant speed U, and look for a steady-state solution in which heat losses and heat conduction normal to the plates are neglected. The problem is then two-dimensional, and non-dimensional space-variables x and y are defined with their origin on the axis of the laser beam, so that the metal sheets are moving past the laser as in Fig. Q1. The laser may be surrounded by a sheath in which case it may be modelled as a hot solid body; with no sheath it is a region of vaporized metal which cannot sustain shear. If the stream function is made non-dimensional with aU and the non-dimensional temperature $u = (T - T_m)/(T_m - T_0)$, then u satisfies

$$q\frac{\partial u}{\partial x} = \frac{\partial^2 u}{\partial x^2} + \frac{\partial^2 u}{\partial y^2} \quad \text{in the solid metal,} \qquad (Q1)$$

$$q\left(\frac{\partial \psi}{\partial y}\frac{\partial u}{\partial x} - \frac{\partial \psi}{\partial x}\frac{\partial u}{\partial y}\right) = \frac{\partial^2 u}{\partial x^2} + \frac{\partial^2 u}{\partial y^2} \quad \text{in the molten metal,} \qquad (Q2)$$

where $q = \rho c U a / k$ is the Peclet number.

If the Reynolds number Ua/ν is small, then the molten metal is modelled as a slow viscous fluid and $\nabla^4 \psi = 0$. The boundary conditions on the laser may be taken as $u = u_1 > 0$, $\psi = 0$ and either $\partial \psi/\partial r = 0$ (no slip) or $\sigma_{r\theta} = 0$ (no shear) on $r = 1$; the conditions far from the laser are $r \to \infty$, $u \to -1$. On the phase-change boundary, defined by $r = f(\theta)$, $u = 0$ and the Stefan condition of eqn (4.14) takes the form

$$[\nabla u]_s^l \cdot \nabla[r - f(\theta)] = q\lambda \frac{\partial}{\partial x}[r - f(\theta)], \qquad (Q3)$$

where the time derivative has been replaced by the convective derivative and λ is the Stefan number $L/c(T_m - T_0)$. The fluid velocity also has to be continuous across the phase-change boundary, assuming no density change between the two phases, so that $\partial\psi/\partial r = \sin\theta$, $(1/r)\partial\psi/\partial\theta = \cos\theta$.

This is a complicated boundary-value problem, and we consider the case $q \ll 1$ when conduction dominates convection. In this case the melt-pool is nearly circular, so that $f(\theta) = R[1 + O(q)]$ and the Stefan boundary conditions may be evaluated on $r = R$.

The flow problem in the no-slip case may be solved by writing $\nabla^4\psi = 0$ in polar coordinates, and looking for a solution in the form $\psi = F(r)\sin\theta$ where $F(1) = F'(1) = 0$, $F(R) = R$, and $F'(R) = 1$. After some manipulation the result is

$$F(R) = \left(\frac{R^2+1}{R^2-1}\log R - 1\right)^{-1}$$

$$\times\left(-\frac{r}{2} + \frac{R^2+1}{R^2-1}r\log r + \frac{R^2}{2r(R^2-1)} - \frac{r^3}{2(R^2-1)}\right). \quad (Q4)$$

Fortunately in this approximation we may neglect the convection terms in eqn (Q1), and we have to solve

$$\frac{\partial^2 u}{\partial r^2} + \frac{1}{r}\frac{\partial u}{\partial r} + \frac{1}{r^2}\frac{\partial^2 u}{\partial \theta^2} = O(q), \quad 1 < r < R,$$

with $u = u_1$ on $r = 1$ and $u = 0$ on $r = R$. The solution is independent of θ, and hence in the molten region

$$u = u_1 - \frac{u_1 \log r}{\log R} + O(q). \quad (Q5)$$

For the solid region $r > R$ the boundary conditions are $r = R$, $u = 0$; $r \to \infty$, $u \to -1$; and the Stefan condition of eqn (Q3) reduces to

$$\left[\frac{\partial u}{\partial r}\right]_s^l = O(q)$$

so that on $r = R$,

$$\frac{\partial u}{\partial r} = -\frac{u_1}{R\log R}$$

is the approximate condition. There is no steady-state solution of the equation (Q1) which satisfies the condition as $r \to \infty$, other than $u \equiv -1$, and to find a solution satisfying the given boundary conditions the convective term in eqn (Q2) must be retained, so that in polar coordinates

$$\frac{\partial^2 u}{\partial r^2} + \frac{1}{r}\frac{\partial u}{\partial r} + \frac{1}{r^2}\frac{\partial^2 u}{\partial \theta^2} = q\left(\cos\theta\,\frac{\partial u}{\partial r} - \frac{\sin\theta}{r}\frac{\partial u}{\partial \theta}\right).$$

We make the transformation $(u+1) = e^{qr\cos\theta}w$ where w satisfies

$$\frac{\partial^2 w}{\partial r^2} + \frac{1}{r}\frac{\partial w}{\partial r} + \frac{1}{r^2}\frac{\partial^2 w}{\partial \theta^2} = q^2 w \tag{Q6}$$

and $w \to 0$ as $r \to \infty$.

This has solutions obtained by separating the variables in the form $w = f_n(r)e^{\pm in\theta}$, where f_n satisfies

$$f_n'' + \frac{1}{r}f_n' - \left(\frac{n^2}{r^2} + q^2\right)f_n = 0; \quad f_n \to 0, r \to \infty.$$

These conditions define the modified Bessel function $K_n(qr)$ (Bessel function of imaginary argument), which has the asymptotic properties $K_n(z) \sim e^{-z}/\sqrt{z}$ as $z \to \infty$ and $K_0(z) \sim \log z + \gamma$ as $z \to 0$, where γ is Euler's constant. Hence

$$w = \sum_{n=0}^{\infty} K_n(qr)\,(C_n \cos n\theta + D_n \sin n\theta)$$

where, to satisfy the condition $w = e^{-qR\cos\theta}$ on $r = R$, C_n and D_n are defined as Fourier coefficients by

$$K_n(qR)C_n = \frac{1}{2\pi}\int_0^{2\pi} e^{-qR\cos\theta}\cos n\theta\,d\theta$$

and

$$K_n(qR)D_n = \frac{1}{2\pi}\int_0^{2\pi} e^{-qR\cos\theta}\sin n\theta\,d\theta = 0. \tag{Q7}$$

The second boundary condition on $r = R$ requires that

$$-\frac{u_1}{R\log R} \sim q\cos\theta + e^{qR\cos\theta}\sum_{n=0}^{\infty} qK_n'(qR)C_n \cos n\theta$$

for small q, where R depends on q and is large. If we expand the right hand side for small q, with qR also small, then from eqn (Q7) the asymptotically largest term in the series is given by $n = 0$ and $C_0 \sim 1/K_0(qR)$. Hence

$$-u \sim \frac{\log R}{\log(qR) + \gamma},$$

using the asymptotic form of K_0 for small argument, and finally

$$R \sim \left(\frac{1}{q}\right)^{u_1/1+u_1} \qquad (Q8)$$

gives the radius of the melt-pool, consistent with the assumption that $Rq \ll 1$. Clearly in the limit $q \to 0$ this molten region extends to infinity and there is no steady state melt-pool without convection. Alternative models with no shear, inviscid molten metal, and temperature-dependent viscosity, are discussed by Dowden, Davis and Kapadia (1983).

It is also interesting to consider the limit in which $q \gg 1$ and convection dominates conduction. The melt-pool will now be a long thin region as shown in Fig. Q2. Upstream of the laser the molten region will be confined to a thin layer in which the flow rapidly changes direction and the governing equations can be simplified. In this upstream molten layer there must be a balance between the convective and diffusion terms, so that the Peclet number based on the layer thickness instead of the circle radius must be O(1). Thus the layer thickness is O(1/q) since q is linear in a, and we define a variable $z = q(r - 1)$ in the layer with the Stefan boundary given by $z = h(\theta')$. The flow problem then reduces to $\partial^4 \psi/\partial z^4 = 0$ if powers of $1/q$ are neglected, and has to

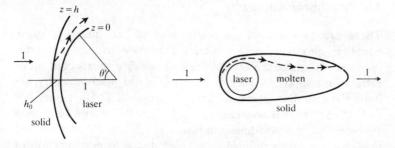

Fig. Q2. Details of flow in a melt-pool

satisfy boundary conditions $\psi = 0 = \partial\psi/\partial z$ on $z = 0$ with $\partial\psi/\partial\theta' = \cos\theta'$ and $\partial\psi/\partial z = O(1/q)$ on $z = h$. Hence

$$\psi = \frac{\sin\theta'}{h^3}(3h - 2z)z^2. \tag{Q9}$$

Equation (Q2) for u reduces to

$$\frac{\partial^2 u}{\partial z^2} = 6\sin\theta'\,\frac{(h-z)z}{h^3}\frac{\partial u}{\partial\theta'} - \cos\theta'\,\frac{z^2(3h-2z)}{h^3}\frac{\partial u}{\partial z}, \tag{Q10}$$

which is a convective diffusion equation with variable coefficients where θ' is the time-like variable. The temperature boundary conditions are

$$z = 0,\, u = u_1; \quad z = h,\, u = 0,$$

together with the Stefan condition of eqn (Q3) which reduces to $\partial u/\partial z = \lambda\cos\theta'$ on $z = h$. It is not possible to integrate eqn (Q10) in closed form and explicitly determine h, but we can derive an expression for the non-dimensional 'stand-off' distance $h(0) = h_0$ be evaluating eqn (Q10) for small z. Since $\partial u/\partial\theta'$ is bounded on $z = 0$, then on this line, where $u = u_0(z)$,

$$\frac{\mathrm{d}^2 u_0}{\mathrm{d}z^2} = -z^2\frac{(3h_0 - 2z)}{h_0^3}\frac{\mathrm{d}u_0}{\mathrm{d}z}.$$

This may be integrated to give

$$\lambda\int_0^{h_0}\exp\left(\frac{z^4}{2h_0^3} - \frac{z^3}{h_0^2} + \frac{h_0}{2}\right)\mathrm{d}z = u_1. \tag{Q11}$$

4.5. Non-linear diffusion

There are two natural generalizations of the linear diffusion equation into non-linear situations, and we expect that non-linearity will introduce interesting and possibly unexpected properties of the solutions, such as occurred in the quasi-linear hyperbolic problem discussed in Chapter 2, Section 2.7. The extensions are either to have a non-linear source term $f(u)$ in eqn (4.1) as in Problem A, or a non-linear diffusion coefficient $D(u)$. Only the case of a non-linear diffusion coefficient in the form of a power of u, possibly fractional, will be discussed here and we consider the *percolation equation*

$$\frac{\partial u}{\partial t} = \frac{\partial}{\partial x}\left(u^n \frac{\partial u}{\partial x}\right) \quad (n \geqslant 0) \tag{4.35}$$

in one space-dimension.

The simplest application is to the percolation of a gas in a porous matrix in which the Darcy law $\boldsymbol{q} = -\kappa \,\mathrm{grad}\, p$ holds, and there is a barotropic gas law $p = p(\rho)$ between the gas pressure p and density ρ. From mass conservation for a gas

$$\frac{\partial \rho}{\partial t} + \mathrm{div}(\rho \boldsymbol{q}) = 0,$$

so that

$$\frac{\partial \rho}{\partial t} = \mathrm{div}[\kappa \rho p'(\rho)\mathrm{grad}\, \rho].$$

For an isothermal gas, p is proportional to ρ and in one space-dimension this is eqn (4.35) with $n = 1$. For an adiabatic gas p is proportional to ρ^γ, where $\gamma > 1$ is the ratio of the specific heats and in eqn (4.35) $n = \gamma$.

A second example, which relies on an empirical law, is the infiltration of moisture into a dry soil. The moisture concentration c is related to the flux of moisture $\boldsymbol{Q}$ by the conservation law $\partial c/\partial t + \mathrm{div}\, \boldsymbol{Q} = 0$, and it is assumed that a pressure potential p exists such that $\boldsymbol{Q} = -\kappa \,\mathrm{grad}\, p$. This pressure potential is the sum of a capillary potential ψ and gravity $-gz$, and both κ and ψ are functions of c which have to be determined from experiment. A power law is commonly used to fit the data, and both κ and ψ may be approximated by a cubic in c. This leads to the *infiltration equation*

$$\frac{\partial c}{\partial t} = \mathrm{div}[a_1 c^3 \,\mathrm{grad}(a_2 c^3 - gz)].$$

In the vertical direction

$$\frac{\partial c}{\partial t} + 3ga_1 c^2 \frac{\partial c}{\partial z} = 3a_1 a_2 \frac{\partial}{\partial z}\left(c^5 \frac{\partial c}{\partial z}\right),$$

a non-linear convective-diffusion equation, but horizontally the percolation equation (4.35) with $n = 5$ is obtained.

We shall discuss only initial-value problems in which $u = u_0 \geqslant 0$ is prescribed at $t = 0$ and u vanishes as $|x| \to \infty$. This corresponds to an initial region or distribution of gas or moisture, and its evolution in time in the two applications above. As in the case of a linear diffusion problem the solution in the non-linear case is never negative, but the situation is different in that the diffusion coefficient (although never negative) can vanish. This will make it possible to find solutions with *compact support*, that is solutions which vanish everywhere outside a range of x which varies with t. We shall therefore be interested in free boundaries $x = s(t)$ beyond which the solution is identically zero, corresponding to the gas and moisture 'fronts' in the two applications above. To define such problems on the whole x-space a weak formulation is required, and an appropriate definition for an initial-value problem is

$$\int_0^{t_0} \int_{-\infty}^{\infty} \left(\frac{u^{n+1}}{n+1} \frac{\partial^2 \phi}{\partial x^2} + u \frac{\partial \phi}{\partial t} \right) dx \, dt + \int_{-\infty}^{\infty} (\phi)_{t=0} u_0(x) \, dx = 0,$$

(4.36)

for all test functions ϕ with continuous first derivatives in t and second derivatives in x which vanish as $|x| \to \infty$ and on $t = t_0$.

It can be shown, as in Elliott and Ockendon (1982), that the weak solution is unique, but existence is not guaranteed for all t. Using Green's theorem it is easily shown, as described earlier for the case $n = 0$ in eqn (4.21), that if u is a classical solution of eqn (4.35) satisfying $u = u_0$ on $t = 0$ and $u \to 0$ as $|x| \to \infty$, then it is also a weak solution. Conversely if u is a weak solution which is positive for all x and t, then it is also a classical solution. If it is a weak solution with $u = 0$ in $x \geqslant s(t)$ and $u > 0$ in $x < s(t)$, then u is a classical solution in $x < s(t)$ and

$$\int u\phi \, dx = \int \left(\frac{u^{n+1}}{n+1} \frac{\partial \phi}{\partial x} - \phi u^n \frac{\partial u}{\partial x} \right) dt,$$

where the integrals are evaluated on $x = s(t)$ for $0 < t < t_0$. Since $u \to 0$ as $x \to s(t)$ from below, this integral identity will be valid for all ϕ provided $u(ds/dt) \to -u^n(\partial u/\partial x)$, that is

$$x = s(t), \quad n \frac{ds}{dt} + \frac{\partial}{\partial x}(u^n) = 0, \quad u = 0 \qquad (4.37)$$

is the free-boundary condition. A free boundary with finite velocity is clearly not possible if $n = 0$, the linear case, but can occur for $n > 0$ if $\partial(u^n)/\partial x$ is bounded and non-zero. When $n = 1$ this implies a profile slope which is finite, but the slope is unbounded for $n > 1$ and zero for $n < 1$. This existence of solutions with compact support for $t > 0$, whose boundaries may be thought of as wavefronts (although the term wavefront is more commonly restricted to hyperbolic problems), is a major qualitative difference between linear and non-linear diffusion.

A second difference is that there is no maximum-value theorem for the non-linear equation, but a comparison theorem is available for the weak solution and may be stated as follows:

If u_1 and u_2 are both solutions of eqn (4.36) (or 4.35) and $0 \leq u_1 \leq u_2$ at $t = 0$ for all x, then $0 \leq u_1 \leq u_2$ for all $t > 0$ *for which they both exist.*

The proof for the classical solution is easily obtained by contradiction. If after some time $u_2 < u_1$, then there must exist a time t_0 at which the profiles of u_1 and u_2 touch, say at x_0, and $\partial u_1/\partial t > \partial u_2/\partial t$ there. Consider

$$f = u_1^n \frac{\partial u_1}{\partial x} - u_2^n \frac{\partial u_2}{\partial x},$$

so that from eqn (4.35) $\partial f/\partial x > 0$ in the neighbourhood of (x_0, t_0) and $f = 0$ at (x_0, t_0). But at $t = t_0$ and near $x = x_0$, $\partial u_1/\partial x < \partial u_2/\partial x$ for $x > x_0$ and $u_1 < u_2$; hence $f < 0$, which gives a contradiction.

The value of the comparison theorem lies in its use with a class of special solutions found by similarity methods, that is by reducing the problem to an ordinary differential equation in a similarity variable depending on both x and t. This technique was used before in the linear case to obtain the Neumann solution of eqn (4.19) to a Stefan problem, and the solution of eqn (4.31) for the flow of a fluid past a heated semi-infinite flat plate in the case of large Peclet number.

We look for a classical solution in the form

$$u = t^a f(\eta), \quad \text{where} \quad \eta = x/t^b \quad (b > 0).$$

Equation (4.35) then reduces to

$$(f^n f')' = t^{2b-1-an}(af - b\eta f'). \tag{4.38}$$

For the linear problem with $n = 0$, we chose $b = \frac{1}{2}$ and a could have any value. With $a = 0$ the solution $\operatorname{erf}(\eta/2)$ used in eqn (4.19) is easily obtained. With $a = -\frac{1}{2}$ the ordinary differential equation integrates to give $f = \exp(-\eta/4)$, and this is the fundamental solution (given by eqn (4.6)) of the diffusion equation obtained earlier by transforms.

With $n \neq 0$ we choose $a = (2b-1)/n$ and the equation again integrates if $b = -a = 1/(2+n)$. This gives

$$f^n = \frac{-n}{n+2}\left(\frac{\eta^2}{2}\right) + \text{constant}$$

which can only represent a solution of interest for $f \geqslant 0$, so that we look for a free boundary $\eta = \alpha$, that is $s = \alpha t^{1/(2+n)}$. The free-boundary condition of eqn (4.37) reduces to

$$\eta = \alpha, \quad f = 0, \quad f^{n-1}f' = \frac{-\alpha}{n+2}$$

which is satisfied by

$$f^n = \frac{n}{n+2}\frac{\alpha^2 - \eta^2}{2}, \quad -\alpha < \eta < \alpha.$$

Thus the fundamental solution of the percolation equation (4.35) is given by

$$u^n = \frac{n}{2(n+2)t}(\alpha^2 t^{2/(2+n)} - x^2), \quad |x| < \alpha t^{1/(n+2)},$$

$$u^n = 0, \quad\quad\quad\quad\quad\quad\quad\quad\quad\quad\quad |x| > \alpha t^{1/(n+2)}. \tag{4.39}$$

This is sketched in Fig. 4.11 at different times t. It may be shown

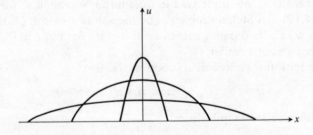

Fig. 4.11. Profiles for point source solution

that this is another representation of the delta function as $t \to 0$ for a suitable value of α.

To avoid algebraic difficulty in the case $n \neq 0$ for more general solutions, we define $F(\eta) = f^n$ so that eqn (4.38) reduces to

$$FF'' + \frac{1}{n}(F')^2 + b\eta F' - (2b-1)F = 0. \tag{4.40}$$

On a free boundary $\eta = \alpha$, $F = 0$ and from eqn (4.37)

$$F' = -nb\alpha \tag{4.41}$$

for any value of $b > 0$. This is a second-order non-linear ordinary differential equation, which can be reduced to autonomous form (no coefficients involving η) by the transformation

$$F = \eta^2 g, \quad F' = \eta h,$$

to give

$$\frac{dh}{dg} = \frac{(g+b)h + (1-2b)g + \frac{1}{n}h^2}{g(2g-h)}, \quad g \geqslant 0. \tag{4.42}$$

Equation (4.42) is a first-order ordinary differential equation, solutions of which may be analysed by the phase-plane method (see, for example, Jordan and Smith (1977)). Given a solution, or an integral curve $h(t)$, then F may be recovered from

$$\log|\eta| = \int^{F/\eta^2} \frac{dg}{h(g) - 2g}. \tag{4.43}$$

There are two critical points of eqn (4.42) in $g \geqslant 0$, namely $(0, 0)$ and $(0, -nb)$, denoted by O and A respectively. Near $(0, 0)$, where both g and h are small, since the denominator of eqn (4.42) is quadratic in g and h then so must be the numerator, and $bh + (1-2b)g \sim 0$. Thus the origin is a saddle-node, and a family of integral curves emanates from the origin; each has slope $dh/dg = 2 - 1/b$, as shown in Fig. 4.9 with $b > \frac{1}{2}$. The point $(0, -n\beta)$ is a saddle point with separatrix of negative slope for all values of $b > \frac{1}{2}$, and a phase-plane portrait may be sketched in as in Fig. 4.12.

For an initial-value problem with non-zero initial data given only on $x < 0$, we require that $t^{2b-1}F(x/t^b)$ is non-zero and

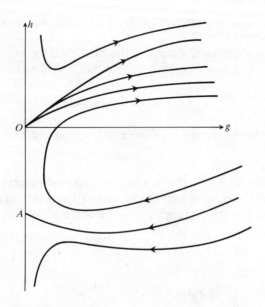

Fig. 4.12. Phase plane for eqn (4.42), $g \geqslant 0$

bounded as $t \to 0$. Hence $F(\eta) \sim |\eta|^{2-1/b}$ as $\eta \to -\infty$, and $g \sim |\eta|^{-1/b}$, $h \sim (2-1/b)|\eta|^{-1/b}$. These are precisely the conditions at the origin of the phase plane, and any integral curve through the origin represents some initial-value problem. It is not difficult to show that these integral curves may be continued smoothly through the point at infinity in the phase plane, where $g \sim kh^2$, to reappear in $h < 0$. This point at infinity corresponds to $\eta \to 0$ and hence to conditions on $x = 0$ in the original variables. One of the integral curves will be continued smoothly into the separatrix of the saddle point at $(O, -b\beta)$. At the saddle point $F'(\eta) = -n\beta\eta$ and $F(\eta) = 0$, and the free-boundary conditions of eqn (4.41) are satisfied with $\eta = ba$. Hence the continuous integral curve from $(0, 0)$ to $(0, -n\beta)$ represents a solution with given initial data of the form

$$u_0 = a\,|x|^{(2b-1)/bn} \quad \text{in} \quad x < 0; \quad u_0 = 0, \quad x > 0,$$

which has a free boundary $x = \alpha t^b$ in $t > 0$, $x > 0$, where α depends on a. Actual values for the solution may be obtained only by numerical integration of eqn (4.42) in the phase plane, followed by the quadrature of eqn (4.43). On $x = 0$, corresponding to f and g large, the form of the solution will be $F \sim kF'^2$, so that $F \sim \eta^2/2k$ and $u = d(x^2/t)^{1/n}$, where d depends on a. For further discussion of phase-plane solutions see Exercise 4.5a.

Problem R. *The spreading of an oil film*

In a number of physical situations it is of interest to predict the spreading of a fluid region on an impermeable planar surface. This may be in the form of a liquid droplet running down a window-pane, when the effects of surface tension will dominate viscous effects except near the liquid boundaries on the plane, or as in the slow spreading of a gas cloud over the ground when compressibility may be important. A simpler problem is to consider a blob of highly viscous oil on a horizontal surface, where the lateral dimensions are large compared to its height. A particular phenomenon which is observed in practice is that an initial blob shape does not necessarily change its planform immediately; the boundaries of the fluid on the horizontal surface may wait a finite time while the vertical shape rearranges itself, after which they then move. It is of interest to construct a model which predicts a waiting time for the front, and obtain conditions for its occurrence.

If the surface height is given by $z = \zeta(x, y, t)$ as in Fig. R1 and we neglect surface tension, then a thin-film approximation can be applied in which z is made non-dimensional with the initial mean height $\bar{\zeta}_0$, but x and y are made non-dimensional with the lateral

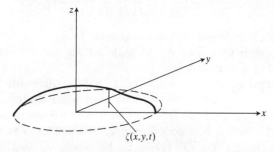

Fig. R1. Oil drop on solid surface

scale $L \gg \bar{\zeta}_0$. This reduces the slow-motion equations (1.24) to the form

$$\frac{\partial p}{\partial x} = \frac{\partial^2 u}{\partial z^2}, \quad \frac{\partial p}{\partial y} = \frac{\partial^2 v}{\partial z^2}, \quad \frac{\partial p}{\partial z} + G = 0 = \frac{\partial u}{\partial x} + \frac{\partial v}{\partial y} + \frac{\partial w}{\partial z},$$

as in the Hele–Shaw cell problem of Chapter 3, Section 3.3, except for the inclusion of gravity G. Here u, v, w are the velocity components with u and v made non-dimensional with $\mu/\rho L$, w with $\mu\bar{\zeta}_0/\rho L^2$, and p with $\mu^2/\rho\bar{\zeta}_0^2$, where $G = \rho^2 g\bar{\zeta}_0^3/\mu^2$ is a non-dimensional parameter which is O(1). The boundary conditions on the horizontal base are

$$z = 0, \quad u = v = w = 0.$$

On the oil surface $z = \zeta$ there is no shear or normal stress, so that

$$p = 0 = \frac{\partial u}{\partial z} = \frac{\partial w}{\partial z} \quad \text{and} \quad w = \frac{\partial \zeta}{\partial t} + u\frac{\partial \zeta}{\partial x} + v\frac{\partial \zeta}{\partial y}, \qquad \text{(R1)}$$

where terms $O(\bar{\zeta}_0/L)$ have been neglected as in Problem M. Integrating with respect to z, $p = G(\zeta - z)$ and

$$u = \frac{G}{2}\frac{\partial \zeta}{\partial x}(z^2 - 2\zeta z), \quad v = \frac{G}{2}\frac{\partial \zeta}{\partial y}(z^2 - 2\zeta z), \qquad \text{(R2)}$$

so that

$$w = -\frac{G}{2}\left\{\left(\frac{\partial^2 \zeta}{\partial x^2} + \frac{\partial^2 \zeta}{\partial y^2}\right)\left(\frac{z^3}{3} - \zeta z^2\right) - z^2\left[\left(\frac{\partial \zeta}{\partial x}\right)^2 + \left(\frac{\partial \zeta}{\partial y}\right)^2\right]\right\}. \qquad \text{(R3)}$$

Substituting eqns (R3) and (R2) in eqn (R1) we obtain

$$\frac{\partial \zeta}{\partial t} = \frac{G}{3}\left[\frac{\partial}{\partial x}\left(\zeta^3\frac{\partial \zeta}{\partial x}\right) + \frac{\partial}{\partial y}\left(\zeta^3\frac{\partial \zeta}{\partial y}\right)\right]. \qquad \text{(R4)}$$

This is the non-linear diffusion equation (4.35) with $n = 3$, if we rescale t with $G/3$. The model with surface tension included is discussed in Exercise 4.5b.

To examine the possibility of a waiting time for the free boundary of the fluid on $z = 0$, we take local coordinates with origin at a point on the free boundary at $t = 0$ and a horizontal scale such that the free boundary may be considered as a straight line; the height is still taken to be small so that the one-dimensional form of eqn (R4) is appropriate. This local problem has initial conditions $\zeta > 0$ for $x < 0$ and $\zeta = 0$ for $x \geqslant 0$ at $t = 0$, so that the similarity solutions $\zeta = t^{(2b-1)/3}[F(\eta)]^{\frac{1}{3}}$ give useful solutions in which $\zeta_0 = c\,|x|^{(2b-1)/3b}$ for $x < 0$. These solutions do not, however, have a waiting time since for $t > 0$ the free boundary is given by $x = \alpha t^b$ where $\alpha > 0$. We therefore look for similarity solutions of this form in which the initial data are given on $t = -t_0 < 0$, and for this purpose we need to extend the similarity transformation to negative values of t. Hence for $t < 0$ define

$$\zeta^n = -(-t)^{2b-1}F(\eta), \quad \text{where} \quad \eta = \frac{-x}{(-t)^b}. \tag{R5}$$

It is easily verified that F satisfies eqn (4.40) above, and the transformation $F = \eta^2 g$, $F' = \eta h$ reduces eqn (4.40) to eqn (4.12) as before, the only difference being that the required integral curves are in the left hand half-space $g < 0$ of the phase plane sketched in Fig. 4.11. We are therefore interested in solutions of eqn (4.42) in the complete g, h space, excluding $g = 0$ where the transformation is not defined.

There is a further critical point B in $g < 0$ at $(-n/2(n+2)$, $-n/(n+2))$, whose character depends on the value of b. Near B, $F \sim [-n/2(n+2)]\eta^2$ and this satisfies the free-boundary condition of eqn (4.41) with $\alpha = 0$, that is $\eta = 0$ is a stationary free boundary for $t < 0$. Hence any integral curve emanating from B with η increasing from zero (corresponding to time increasing to zero from $-t_0$) will lead to a solution with a stationary free boundary at $x = 0$. Analysing the nature of the critical point B it may be shown that it is a node for

$$b > \frac{3n+4}{2(n+2)} + \left[\frac{2n}{n+2}\right]^{\frac{1}{2}},$$

with η increasing from the node. Thus in the case $n = 3$ for $b > 2.4$, $F \sim (-3/10)\eta^2$; and if $\zeta \sim (-3x^2/10t)^{\frac{1}{3}}$ for small values of $x/(-t)^b < 0$, the free boundary $x = 0$ will not move for negative times. In fact $\zeta = (-3x^2/10t)^{\frac{1}{3}}$, $x < 0$; $\zeta = 0$, $x > 0$ is an exact solution of eqn (R4) for $t < 0$ which 'blows up' at $t = 0$. We, however, are looking for solutions which can be continued smoothly into $t > 0$, that is into the region $g > 0$ of the phase plane from the region $g < 0$. An integral curve can only cross the $g = 0$ axis at a critical point, and since A represents a moving free boundary we focus attention on the critical point O.

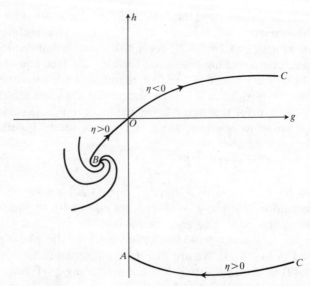

Fig. R2. Phase plane for eqn (4.42), $n = 3$, $b > 2.4$

The critical point O is a saddle node, and it has already been shown that for $g > 0$ a family of integral curves emanate from it, each with initial slope $(2 - 1/b)$. For $g < 0$ there is a unique integral curve into the node which has slope $(2 - 1/b)$, and on it η is increasing. Thus we may hope that an integral curve may be found emanating from B and passing onto O, as in Fig. R2 in the case $n = 3$. Since, near O, $F \sim \gamma |\eta|^{2 - 1/b}$ where γ can be chosen to be the same constant for $g < 0$ and $g > 0$, then as $t \to 0\pm$, $\zeta = a(-x)^{(2b-1)/3b}$ for all $x < 0$. This has derivatives of all orders in x and hence at $t = 0$, $x < 0$; a continuous differentiable solution for ζ may therefore be obtained by following the integral curve BO through O into $g > 0$. There will therefore be a unique trajectory $BOCA$, where C is the point at infinity such that $g \sim kh^2$, which leads to a solution with a stationary free boundary $x = 0$ in $t < 0$, and a moving free boundary in $t > 0$. The waiting time t_0 will be obtained for initial data such that $\zeta = 0$, $x > 0$, and for $x < 0$

$$\zeta_0^3 = -t_0^{2b-1} F(-x t_0^{-b}), \tag{R6}$$

where F depends on $b > 2.4$ and its functional form is to be obtained by evaluating the integral curve BO and then finding

$F(\eta)$ from the quadrature of eqn (4.43). For small values of $x < 0$, this profile will have the asymptotic form $\zeta_0 \sim (3x^2/10t_0)^{\frac{1}{3}}$. It is possible to extend the range of permissible values of b to $b > 2$ when B becomes a spiral point, and for details reference may be made to Lacey, Ockendon and Tayler (1982). Further solutions can also be found which exhibit blow-up.

These similarity solutions may be used with the comparison theorem to obtain bounds on the behaviour of more general initial profiles, and this enables us to establish some properties of the waiting time phenomenon. For the oil blob the boundary will not wait if locally $\zeta_0 > O(x^{\frac{2}{3}})$ and will wait if $\zeta_0 \leqslant O(x^{\frac{2}{3}})$. The actual value of the waiting time will depend on the global properties of $\zeta_0(x)$, but bounds can be obtained for it. The global properties of ζ_0 will also determine whether the boundary moves off smoothly at the end of the waiting time, or whether the solution then blows up. However, oil blob blow-up has not been observed and would appear to be unlikely to occur.

Exercises

4.1. (a) Show that for the boundary-value problem

$$\frac{\partial \phi}{\partial t} = \frac{\partial^2 \phi}{\partial x^2}, \quad -\infty < x < \infty, \quad t > 0; \quad \phi \to 0, \quad x \to \pm\infty;$$

$$\phi = g(x), \quad t = 0, \quad -\infty < x < \infty, \quad \text{where} \quad g \to 0 \quad \text{as} \quad x \to \pm\infty;$$

the Laplace transform of ϕ is given by

$$\hat{\phi}(p, x) = \frac{1}{2\sqrt{p}} \int_{-\infty}^{x} g(s) e^{(x-s)\sqrt{p}} \, ds + \frac{1}{2\sqrt{p}} \int_{x}^{\infty} g(s) e^{-(x-s)\sqrt{p}} \, ds,$$

where the appropriate branch of $\sqrt{p}$ has been chosen. Using the inversion theorem, with a branch cut along the negative real axis and interchanging the order of integration, derive the result of eqn (4.8) for ϕ.

(b) In Problem N, solve the boundary-value problem of eqn (N6) for $\bar{c}$ when $\partial \bar{c}/\partial z = 0$ on $z = d$, to give

$$\bar{c}_{z=0} = \frac{1}{2\pi i} \int_{\Gamma} \frac{e^{px} \cosh[p^{\frac{1}{2}}(d-1)]}{p^{\frac{1}{2}} \sinh(p^{\frac{1}{2}} d)} \, dp.$$

Show that

$$\bar{c}_{z=0} = \frac{1}{d} + O(e^{-\pi^2 x/d^2}).$$

(c) Reconsider Problem N when $U = z^m$ and $D = z^n$ in the case $d \to \infty$. Show that the equation for $\hat{c}$ may be simplified using the transformation $z = s^a$, $\hat{c} = s^r \hat{C}$ to give

$$\frac{d^2 \hat{C}}{ds^2} + \frac{1}{s}\frac{d\hat{C}}{ds} - \left(pq^2 + \frac{r^2}{s^2}\right)\hat{C} = -q\delta(s-1),$$

if $q = 2/(m-n+2)$ and $r = (1-n)/(m-n+2)$. Hence complete the solution in terms of Bessel functions of imaginary argument. Discuss also the special case $m - n + 2 = 0$.

4.2. (a) A solid phase is separated from a liquid phase by the boundary $x = s(t, y) = Vt + \varepsilon e^{\sigma t} \sin ny$, with the solid in $x < s$. Show that for $\varepsilon \ll 1$ a solution is possible to the two-phase Stefan problem correct to $O(\varepsilon)$ in the form

$$T - T_m = A(e^{V(Vt-x)} - 1) + a\varepsilon \sin ny \, e^{\sigma t + m_1(x-Vt)}, \quad x < s;$$
$$= B(e^{V(Vt-x)} - 1) + b\varepsilon \sin ny \, e^{\sigma t + m_2(x-Vt)}, \quad x > s,$$

if $B - A = L$, $a = AV$, $b = BV$ and

$$L\sigma = m_1 a + AV^2 - m_2 b - BV^2,$$

where m_1 and m_2 are the appropriately chosen roots of

$$m^2 + mV - n^2 - \sigma = 0.$$

With $B/A = \mu$, show that

$$(\mu - 1)^2 \sigma = 2\mu V^2 - (\mu + 1)V[\mu V^2 + (\mu - 1)^2 n^2]^{\frac{1}{2}},$$

and hence deduce a stability result for the moving boundary.

(b) Consider the microscopic model for the mushy region described in Problem O when the grain boundaries are planar and distance ε apart. If the grain boundaries near $x = x_0$ are given by $\xi = -2, -1, 0, 1, 2, \ldots$ where $x = x_0 + \varepsilon \xi$, show that in $\sigma_n < \xi < \sigma_{n+1}$

$$\bar{u} = \frac{q}{2}(\xi - \sigma_n)(\sigma_{n+1} - \xi),$$

where the region is solid if n is odd, liquid if n is even. Apply the Stefan condition at $\xi = \sigma_n$ and σ_{n+1} to obtain a difference-differential equation for σ_n. Show that it has a solution

$$\sigma_n = \frac{q}{2\lambda}[n - t + t_1(x_0)] \quad \text{for } n \text{ even;}$$

$$\sigma_n = \frac{q}{2\lambda}[n - 1 + t - t_1(x_0)] \quad \text{for } n \text{ odd.}$$

Verify that the average enthalpy is $q[t - t_1(x_0)]$ and hence recover eqn (O3).

4.3. (a) Look for a similarity solution of eqn (4.23) subject to eqn (4.24) and suitable conditions on $x = 0$ and $t = 0$, in the case $D = 0$ in the solid, in terms of $\eta = x/2\sqrt{t}$ and a phase change boundary $\eta = \alpha$. Show that in $\eta > \alpha$

$$c - c_\infty = (kc_l - c_\infty)\frac{\text{erfc}(\eta/\sqrt{D})}{\text{erfc}(\alpha/\sqrt{D})},$$

where $k = k_s/k_l < 1$, $c_l = -k_l u_m$, and α is to be determined from

$$c_\infty = c_l[k + (1 - k)f(\alpha/\sqrt{D})],$$

$$u_0 - u_\infty + \frac{u_m - u_0}{\text{erf } \alpha} = -\lambda f(\alpha).$$

Here λ is the non-dimensional latent heat and

$$f(\alpha) = \pi^{\frac{1}{2}}\alpha \exp(\alpha^2)\text{erfc}(\alpha).$$

Given that $0 < f(\alpha) < 1$ for all $\alpha \geqslant 0$, and $f(\alpha) \simeq 1$ for large α, show that for small D

$$\frac{c_\infty}{k_l} + u_0 = [\lambda f(\alpha) + u_0 - u_\infty]\text{erf } \alpha$$

and hence deduce that there is not necessarily a solution for α.

(b) A saturated vapour is at rest at temperature u_∞ adjacent to a plane surface $y = 0$. At time $t = 0$ the surface is cooled so that $u = u_w(t) < u_\infty$ on $y = 0$ for $t > 0$. If the vapour condenses in a thin film $0 < y < \delta(t)$, where $\delta \ll 1$ since the fluid density is much greater than that in the vapour, show that the temperature distribution in the film is given by

$$u = u_w(t) + \frac{y}{\delta(t)}[u_p(t) - u_w(t)],$$

where u_p is the phase change temperature, and the concentration $c = -k_l u_p(t)$; use a phase diagram similar to Fig. 4.4b. In the vapour region $y > \delta$ (assumed incompressible) show that, neglecting terms of $O(\delta)$,

$$u = u_\infty + \frac{1}{2\pi} \int_0^t [u_p(s) - u_\infty] \frac{\exp[-y^2/4(t-s)]}{(t-s)^{\frac{1}{2}}} \, ds,$$

and obtain a similar expression for c. From the Stefan conditions of eqn (4.24) write down two equations for $\delta(t)$ and $u_p(t)$ (being careful to avoid divergent integrals). Show that the approximation is only valid if $ku_\infty/\rho\lambda D$ is $O(1)$ where k is the heat conductivity, ρ the density, D the diffusion coefficient in the vapour and λ the latent heat of vaporization.

4.4. (a) A heat source of strength unity at $x = 0$ is switched on at time $t = 0$. An infinite solid bar is moved with speed df/dt past the source, where f increases monotonically. Show that the temperature distribution in the bar at time t is given by

$$\frac{1}{2\sqrt{\pi}} \int_0^t \exp\{-[x + f(s) - f(t)]^2/4(t-s)\} \frac{ds}{(t-s)^{\frac{1}{2}}}.$$

(b) Show that in Problem P if $\varepsilon \ll \delta \ll 1$ the dendritic tip region is given by the scaling $\varepsilon^2 \bar{z} = \delta z$, $\delta y = \varepsilon \bar{y}$ and $\delta f = \varepsilon \bar{f}$, so that the boundary-value problem for the concentration c is

$$\frac{\partial c}{\partial \bar{z}} = \frac{\partial^2 c}{\partial \bar{z}^2} + \frac{\partial^2 c}{\partial \bar{y}^2}, \quad \bar{f} < \bar{y} < \infty, \quad -\infty < \bar{z} < \infty;$$

$$c = -k_l u_0, \quad \frac{\partial c}{\partial \bar{y}} - \frac{d\bar{f}}{d\bar{z}}\frac{\partial c}{\partial \bar{z}} = (k_l - k_s)u_0\frac{d\bar{f}}{d\bar{z}} \quad \text{on} \quad \bar{y} = \bar{f};$$

together with $c = c_\infty$ as $\bar{z} \to -\infty$, $\partial c/\partial \bar{y} = 0$ as $\bar{y} \to \infty$, and $\bar{f} \equiv 0$ for $\bar{z} < 0$. Rewrite this problem in terms of variables $\bar{z} - \eta_0^2 = \xi^2 - \eta^2$,

$\bar{y} = 2\xi\eta$, similar to those used to solve eqn (4.28), to give

$$\frac{d^2c}{d\eta^2} + 2\eta\frac{dc}{d\eta} = 0, \quad \eta \geqslant \eta_0; \quad c = c_\infty, \quad \eta \to \infty;$$

$$c = -k_l u_0, \quad \frac{dc}{d\eta} = \frac{(k_l - k_s)u_0}{2\eta_0} \quad \text{on} \quad \eta = \eta_0; \quad \text{where} \quad \bar{f} = 2(\bar{z}\eta_0)^{\frac{1}{2}}.$$

Show that there exists a unique η_0 given by the root of

$$(k_l - k_s)\mathrm{erfc}(\eta_0) = 4\pi^{-\frac{1}{2}}\left(1 + \frac{c_\infty}{k_l u_0}\right)\eta_0 \exp(-\eta_0^2),$$

provided $c_\infty > -k_l u_0$, that is the tip is undercooled.

4.5. (a) Show that the two (space) dimensional percolation equation with radial symmetry has similarity solutions in the form $u^n = t^{2b-1}F(\eta)$ where $\eta = r/t^b$. If $F = \eta^2 g$ and $F' = \eta h$ show that

$$\frac{dh}{dg} = \frac{(2g+b)h + (1-2b)g + \dfrac{1}{n}h^2}{g(2g-h)}.$$

Verify that the critical points in $g \geqslant 0$ are the same as in the one (space) dimensional case. Evaluate the point-source solution by choosing $b = 1/2(1+n)$ and the integral curve $h = -n/2(1+n)$.

In the one-dimensional case show that there are exact solutions when (i) $b = 1$; (ii) $b = 1/2(1+n)$; (iii) $b = (n+1)/(n+2)$, and describe their properties.

(b) If surface tension is included in Problem R so that the surface-pressure condition is $p = \lambda(\partial^2\zeta/\partial x^2)$, show that the one-dimensional model equation (R4) becomes

$$\frac{\partial\zeta}{\partial t} = \frac{1}{3}\frac{\partial}{\partial x}\left[\zeta^3\left(\lambda\frac{\partial^3\zeta}{\partial x^3} + G\frac{\partial\zeta}{\partial x}\right)\right].$$

Suggest *two* suitable boundary conditions at the free boundary $x = s(t)$. Show that the centre of gravity of the blob does not move in the x direction.

(c) A volcano is assumed to consist of porous material through which molten magma flows from a feeder pipe at the centre of its base. The magma solidifies at the surface, thus increasing the size of the volcano. Using a thin-layer approximation, so that a mean radial velocity u can be defined, show that the volcano surface $h(r, t)$ is given by

$$r \frac{\partial h}{\partial t} = -\frac{\partial}{\partial r}(rhu).$$

If the pressure is hydrostatic in the volcano and Darcy's law holds show that $u = -K(\partial h/\partial r)$, where K depends on material properties, and obtain an equation for h. Show also that at the edge of the volcano base, where $r = s(t)$ and $h = 0$, $\dot{s} = -K(\partial h/\partial r)$.

Look for a similarity solution in terms of $\eta = r/(Kt)^{\frac{1}{2}}$ and $s = \eta_0(Kt)^{\frac{1}{2}}$ and show that $h = f(\eta)$ satisfies

$$(ff')' + \frac{1}{\eta} ff' + \tfrac{1}{2}\eta f' = 0.$$

Obtain suitable boundary conditions at $\eta = \eta_0$ and $\eta = 0$, and discuss the solution of the problem in terms of a phase plane similar to that described for eqn (4.40).

5 Asymptotic methods for non-linear problems

5.1. Non-linear ordinary differential equations

In previous chapters nearly all the partial differential equations modelling the physical phenomena were linear, or they reduced to linear equations on applying a regular perturbation in some small parameter. For linear problems there is a considerable body of mathematical analysis available, so that in almost all cases it was possible to demonstrate clearly that the model was well posed. In Sections 2.7 and 4.5 two types of non-linear equation were discussed, and sufficient analytical results are available to again give us reasonable confidence that the models are well posed. However, almost no mathematical results are available about the well-posedness of any genuinely non-linear boundary-value problem and in developing non-linear models we must proceed with considerable caution.

One way of testing such a model is to examine it for small values of one or more of the parameters in the problem, to check whether for these values the problem is well posed; such a solution (if it can be found) may also be used to check any proposed numerical procedures. However (as was briefly discussed in Chapter 1) it may not be possible to obtain the asymptotic form of the problem by using a regular expansion in the small parameter ε, and the exact solution for $\varepsilon = 0$ may be fundamentally different in character from the 'neighbouring' solutions obtained in the limit $\varepsilon \to 0$. In such a situation, which frequently occurs, singular-perturbation techniques are required. These techniques are somewhat *ad hoc*, that is appropriate to a specific problem and difficult to categorize; moreover, few mathematical proofs are available to show that they are truly asymptotic. They are rational rather than rigorous procedures, which may be confirmed by subsequent computation or by the success of the interpretation of the asymptotic solution to the model in its applications.

It is, however, possible to describe singular-perturbation techniques in three or four overlapping main types, and to illustrate them simply we first consider in each case an elementary problem involving an ordinary differential equation, before attempting to apply these techniques to partial differential equations in the remaining sections of this chapter. For a detailed discussion of singular perturbation techniques applied to ordinary differential equations see Chapters 7 and 9–11 of Bender and Orszag (1978).

(a) *The WKB method.* This has already been described briefly in Chapter 1, in the context of finding asymptotic solutions of a second-order ordinary differential equation for large values of the independent variable. Here we look for asymptotic solutions for small ε of

$$\varepsilon^2 \frac{d^2 u}{dx^2} + h(x, \varepsilon)u = 0, \qquad (5.1)$$

where h is regular in ε.

Assuming an asymptotic expansion in the form

$$u = \exp \int^x \left[\frac{1}{\varepsilon} g_0(s) + g_1(s) + \varepsilon g_2(s) + \ldots \right] ds$$

for u, and the regular expansion $h = h_0(x) + \varepsilon h_1(x) + \ldots$, we equate powers of ε on substitution into eqn (5.1). Then

$$g_0^2 + h_0 = 0, \quad g_0' + 2g_0 g_1 + h_1 = 0, \quad \text{etc.}$$

Hence

$$g_0 = (-h_0)^{\frac{1}{2}}, \quad g_1 = \frac{-h_0'}{4h_0} - \frac{h_1}{2(-h_0)^{\frac{1}{2}}}, \quad g_2 = \ldots,$$

and

$$u \sim \frac{\exp \int^x \frac{1}{\varepsilon} [-h_0(s)]^{\frac{1}{2}} \left(1 + \varepsilon \frac{h_1(s)}{2h_0(s)} + \ldots \right) ds}{[-h_0(x)]^{\frac{1}{4}}}. \qquad (5.2)$$

Clearly this is not a regular expansion in ε, since the leading term is $O(e^{\pm 1/\varepsilon})$ and it can never be asymptotic to a power of ε for small ε. We expect a singular perturbation in eqn (5.1) because with $\varepsilon = 0$ the problem reduces to $h(x, 0)u = 0$ and *the highest-order derivative disappears.* Thus if we had attempted to use a regular expansion every term would have been zero, and no boundary conditions could have been satisfied other than the trivial ones $u = du/dx = 0$.

The validity of eqn (5.2) as an asymptotic expansion is not easy to prove, and it is not valid for all values of h. This is fairly obvious, since in carrying out the expansion procedure it has been implicitly assumed that g_0, g_1, etc. are bounded functions; near a zero of $h_0(x)$, where g_1 is singular, we must expect that the procedure will not be uniformly valid. The transformation $x \to \varepsilon^n z$ reduces eqn (5.1) to

$$\frac{d^2 u}{dz^2} + \varepsilon^{2n-2} h(\varepsilon^n z, \varepsilon) u = 0,$$

and in the special case when $h(z) = \varepsilon^{2n-2} h(\varepsilon^n z, \varepsilon)$ we recover the problem discussed in Chapter 1 and the asymptotic result of eqn (1.21). In this context the zeros of $h(z)$ where eqn (1.21) is not valid are called *transition points.*

The WKB method is not restricted to second-order equations, but it does require the problem to be linear. A novel example occurs in the solution of Problem B in Chapter 2 where, following eqn (B8), we wish to find the solution for $\varepsilon \ll 1$ of

$$\frac{du}{dt} = k_1 u[(1-\varepsilon)t] + k_2 u(t), \quad u(0) = 1.$$

A regular expansion in powers of ε leads to

$$u_0 = e^{(k_1+k_2)t}, \quad 2u_1 = -k_1(k_1 + k_2)t^2 e^{(k_1+k_2)t}, \quad \text{etc.}$$

For large t, $\varepsilon u_1/u_0 \sim \varepsilon t^2$ and the nth-order term will not be uniformly smaller than the $(n-1)$th-order term when $t \sim 1/\varepsilon$. Thus the expansion is singular for large t, and we rescale the problem by writing $\tau = \varepsilon t$ and $v(\tau) = u(\tau/\varepsilon)$ to examine this range of variables in detail.

Equation (5.3) becomes

$$\varepsilon \frac{dv}{d\tau} = k_1 v(\tau - \varepsilon\tau) + k_2 v(\tau),$$

and in these variables a regular expansion is not possible since with $\varepsilon = 0$ the highest derivative disappears. It is, however, a linear problem so we attempt a WKB expansion in the form

$$v = \exp \int_0^\tau \left[\frac{g_0(s)}{\varepsilon} + g_1(s) + O(\varepsilon) \right] ds$$

where $v(0) = u(0) = 1$ is satisfied. Substituting in the equation, g_0 is determined from

$$g_0(\tau) = k_1 \exp \int_\tau^{\tau - \varepsilon\tau} \frac{g_0(s)\, ds}{\varepsilon} + k_2 + O(\varepsilon)$$

$$= k_1 e^{-\tau g_0(\tau)} + k_2, \qquad (5.4)$$

where terms of $O(\varepsilon)$ are put equal to zero by a suitable choice of g_1, etc. It may be verified that $g_i(\tau)$ is an asymptotic sequence for all τ so that we have found a uniformly valid asymptotic form for v, and hence u, provided that eqn (5.4) has bounded solutions for g_0. The solution of eqn (5.4) for various ranges of the parameters k_1 and k_2 is sketched in Fig. 5.1; for $k_1 > 0$, g_0 is a monotonic function of τ which is positive if $k_1 + k_2 > 0$. This implies that v increases with τ, and that solutions of eqn (5.3) are unbounded as $t \to \infty$. With $k_1 + k_2 < 0$ they decay to zero. If $k_1 < 0$ and $k_1 + k_2 > 0$ there are two possible real roots for g_0, one positive and one negative; for $k_1 < 0$ and $k_1 + k_2 < 0$ there are no real roots for g_0 for a range of values of τ. The implications of this non-uniformity

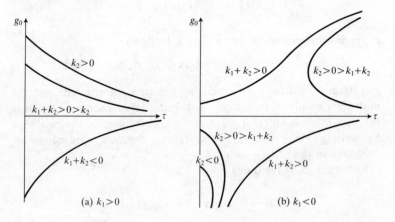

Fig. 5.1. Solution curves for eqn (5.4)

in g_0 on the solutions of eqn (5.3) for large time are discussed further in Exercise 5.1a.

(b) *Matched expansions.* Singular perturbations, which occur because the term containing the highest derivative disappears when $\varepsilon = 0$, will give rise to isolated regions of non-uniformity. These regions usually (but not necessarily) occur at the boundary of the domain, and are often called *boundary layers.* In the boundary layer the term containing the highest derivative is no longer small, because the highest derivative is large. To avoid even larger changes in the function itself the layer has to be thin. An appropriate new variable is chosen which 'magnifies' or stretches the boundary layer, and leads to an 'inner' problem for the solution within the layer. If this can be solved in terms of simple functions then the solution has to be 'matched' to the 'outer' solution, obtained by a regular perturbation expansion in the original variables and valid outside the boundary layer.

The skill needed to use the method lies in locating the boundary layer or layers, and in choosing appropriate inner variables. If we are content to obtain only the first term in the asymptotic expansions (the zero-order solution), then the matching procedure simply requires continuity of the inner and outer solutions between their two regions of validity. However, to correctly match more than the first term requires considerable care and the use of a matching procedure, as described in Van Dyke (1964).

A simple example illustrates the salient features; consider

$$\varepsilon \frac{d^2 u}{dx^2} = \frac{du}{dx} + 1 + \varepsilon c u^2; \quad u = 0, \quad x = 0 \text{ and } 1. \quad (5.5)$$

The outer expansion is $u = a - x + O(\varepsilon)$, where a is a constant to be determined. Clearly both boundary conditions cannot be satisfied by one choice of a, and there must be a non-uniform region where $d^2 u/dx^2$ is large. Let us assume that such a region occurs adjacent to $x = 0$; then stretching this layer by the transformation $x = \varepsilon^n z$ and writing $u(x) = v(z)$, eqn (5.5) becomes

$$\varepsilon^{1-n} \frac{d^2 v}{dz^2} = \frac{dv}{dz} + \varepsilon^n + \varepsilon^{1+n} c v^2.$$

The only choice of n which retains the highest derivative and one other term is $n = 1$, and with $v = v_0(z) + \varepsilon v_1(z) + \ldots$ and $v(0) = 0$

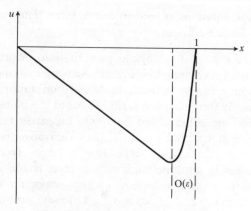

Fig. 5.2. Solution curve for eqn (5.5)

we obtain

$$v_0 = A(e^z - 1).$$

The matching condition requires that v_0 is bounded for large z, that is for $O(1)$ values of x and hence $A = 0$. There is therefore no boundary layer at $x = 0$, and we must apply the boundary condition there to give $a = 0$.

If we now assume a boundary layer at $x = 1$, then an appropriate new variable is given by $1 - x = \varepsilon z$ and

$$\frac{d^2v}{dz^2} = -\frac{dv}{dz} + \varepsilon + \varepsilon^2 c v^2, \quad v(0) = 0.$$

In this case $v_0 = A(e^{-z} - 1)$, and the matching condition that v is continuous for large z with u gives $-A = u(1) = -1$. The zero-order solution is sketched in Fig. 5.2, and can be confirmed by an exact solution in the case $c = 0$. It is conceivable that there could be an internal boundary layer near $x = x^*$, and to investigate this a boundary-layer variable $z = (x^* - x)/\varepsilon$ would be appropriate. The matching would not, however, be possible for large negative z for the same reason that we rejected the layer at $x = 0$. There is no difficulty in obtaining some further terms in these expansions, and correct to $O(\varepsilon)$ they are

$$\text{outer:} \quad u = -x - \frac{\varepsilon c x^3}{3},$$

$$\text{inner:} \quad v = e^{-z} - 1 + \varepsilon z + \varepsilon B(e^{-z} - 1),$$

where B is to be determined from the matching For large z, the inner solution $v \sim -1 + \varepsilon(z - B) + O(\varepsilon^2)$. Hence we choose $B = c/3$ to give matching correct to $O(\varepsilon)$.

In more complicated problems it may be necessary to rescale the dependent variable in addition to choosing a boundary layer variable, and the appropriate inner expansion may well be in terms of an asymptotic sequence which is not a power series as in Exercise 5.1b. Some skill and a process of trial and error are often both needed to make these delicate choices, and there are almost no general rules to be followed. The success of the method depends crucially on there being a closed-form solution to the zero-order inner and outer problems, although the location of a boundary layer and an estimate of its thickness may by itself be valuable information, especially if a numerical solution is contemplated.

(c) *Two-timing and multiple scales.* A non-uniformity in a regular expansion may also arise as a result of a resonant interaction between different terms in the expansion. An example of this is the Mathieu equation (3.17) which was used to model the surface disturbance of an inviscid fluid oscillating in a vertical direction in a tank. In the notation of this section the problem is

$$\frac{d^2u}{dt^2} + (n^2 - \varepsilon \cos t)u = 0, \tag{5.6}$$

with boundary conditions at $t = 0$, say $u = 1$, $du/dt = 0$. If $\varepsilon = 0$, $u = \cos nt$, and if $\varepsilon \ll 1$ with $u = u_0 + \varepsilon u_1 + \ldots$, then $u_0 = \cos nt$ and

$$\frac{d^2u_1}{dt^2} + n^2 u_1 = \cos nt \cos t = \tfrac{1}{2}\{\cos[(1+n)t] + \cos[(1-n)t]\}.$$

This has bounded solutions for u_1 for all t, provided that $(1+n)$ and $(1-n)$ are both not equal to n. Thus when $n = \tfrac{1}{2}$ the solution for u_1 contains the *secular* term $\tfrac{1}{2}t \sin(t/2)$, and the expansion may be expected to be valid only for times t such that $\varepsilon t \ll 1$. There are two time scales in the problem, one for the fundamental oscillation and one for the effect of the perturbation terms. As in the case of eqn (5.3) we would expect to have to introduce the time scale $\tau = \varepsilon t$ into the problem, but for eqn (5.6) both time scales are equally important and we cannot eliminate t. Thus we are led to introduce two times t and τ, and the method of 'two-timing' rewrites the problem as a partial differential equation in the variables t and τ. With $u = u(t, \tau)$, so that $\partial/\partial t$ is replaced by $(\partial/\partial t + \varepsilon(\partial/\partial \tau))$, eqn (5.6) becomes

$$\frac{\partial^2 u}{\partial t^2} + 2\varepsilon \frac{\partial^2 u}{\partial t\, \partial \tau} + \varepsilon^2 \frac{\partial^2 u}{\partial \tau^2} + (n^2 - \varepsilon \cos t) u = 0.$$

We consider values of n close to $\frac{1}{2}$, (where the non-uniformity occurs) by writing $n^2 = \frac{1}{4} + \lambda \varepsilon$, and we expand u in powers of ε. Then substituting in the equation and equating powers of ε to zero

$$\frac{\partial^2 u_0}{\partial t^2} + \frac{u_0}{4} = 0,$$

$$\frac{\partial^2 u_1}{\partial t^2} + \frac{u_1}{4} = u_0(\cos t - \lambda) - 2 \frac{\partial^2 u_0}{\partial t\, \partial \tau}, \quad \text{etc.}$$

Hence

$$u_0 = a(\tau)\cos \frac{t}{2} + b(\tau)\sin \frac{t}{2},$$

where $a(0) = 1$, $b(0) = 0$ to satisfy the boundary conditions. Thus

$$\frac{\partial^2 u_1}{\partial t^2} + \frac{u_1}{4} = a \cos t \cos \frac{t}{2} + b \cos t \sin \frac{t}{2}$$

$$+ \left(\frac{\mathrm{d}a}{\mathrm{d}\tau} - \lambda b\right)\sin \frac{t}{2} - \left(\frac{\mathrm{d}b}{\mathrm{d}\tau} + \lambda a\right)\cos \frac{t}{2}$$

$$= F(t).$$

We want to choose $a(\tau)$ and $b(\tau)$ so that u_1 is bounded for all t where $F(t)$ is a periodic function, period 4π. Now $F(t)$ can be expanded in a Fourier series

$$\sum_{r=0}^{\infty} a_r \cos \frac{rt}{2} + \sum_{r=1}^{\infty} b_r \sin \frac{rt}{2},$$

and secular terms will occur in the solution for u_1 if either a_1 or b_1 are non-zero. Thus the necessary condition for u_1 to be bounded is that

$$\int_0^{4\pi} F(t)\cos \frac{t}{2}\, \mathrm{d}t = 0 = \int_0^{4\pi} F(t)\sin \frac{t}{2}\, \mathrm{d}t.$$

Substituting for $F(t)$ and evaluating the integrals we obtain

$$\frac{\mathrm{d}a}{\mathrm{d}\tau} - (\lambda + \tfrac{1}{2})b = 0 = \frac{\mathrm{d}b}{\mathrm{d}\tau} + (\lambda - \tfrac{1}{2})a,$$

so that

$$\frac{\mathrm{d}^2 a}{\mathrm{d}\tau^2} + (\lambda^2 - \tfrac{1}{4})a = 0.$$

If $|\lambda| > \tfrac{1}{2}$ this has periodic solutions, and

$$u = \cos[\varepsilon(\lambda^2 - \tfrac{1}{4})^{\frac{1}{2}}t]\cos\frac{t}{2} + \sin[\varepsilon(\lambda^2 - \tfrac{1}{4})^{\frac{1}{2}}t]\sin\frac{t}{2} + \mathrm{O}(\varepsilon) \quad (5.7)$$

uniformly for all t.

In the context of Fig. 3.2 we have found the stable regions of the solution in the neighbourhood of $\varepsilon = 0$, $n = \tfrac{1}{2}$. The shaded unstable region will be given by $|\lambda| < \tfrac{1}{2}$, that is $|4n^2 - 1| < \varepsilon$, where we have found a uniformly valid solution which grows exponentially in τ.

Resonances will occur between higher-order terms and the zero-order term in any regular expansion of eqn (5.6). Thus

$$\frac{\mathrm{d}^2 u_2}{\mathrm{d}t^2} + n^2 u_2 = u_1 \cos t,$$

and the right hand side contains terms proportional to $\cos nt$, $\cos[(2+n)t]$ and $\cos[(2-n)t]$. This will give rise to a secular term if $n = 2 - n$, that is $n = 1$; similarly subsequent higher-order terms will resonate at half-integer values of n. For the resonance at $n = 1$, the choice of the second time scale τ would now be $\varepsilon^2 t$, since $\varepsilon^2 u_2/u_0 \sim \varepsilon^2 t$.

The two-timing method may be applied to a variety of oscillatory problems, and the choice of the second time scale is not always obvious; see, for example, Chapter 11 of Bender and Orszag (1978) and Exercise 5.1c. It may also be generalized to non-oscillatory problems, where it is usually called the *method of multiple scales*; this embodies in addition the ideas of boundary layer and WKB theory. There is a considerable literature, consisting mostly of examples rather than theory, and reference may be made to Jeffrey and Kawahara (1982) and others.

Problem S. Resonance in a gearbox

The primary function of a gearbox is to effect a change in angular speed and to transmit power from the driving mechanism, such as a motor or turbine, to the mechanism being driven (the output device). We may represent a simple gearbox schematically as in

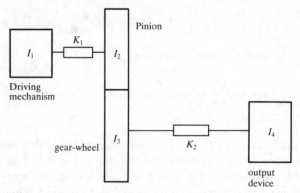

Fig. S1. Schematic diagram of gear box

Fig. S1. If the driving mechanism has moment of inertia I_1, and drives a pinion with moment of inertia I_2, through a coupling modelled by a torsional stiffness K_1, then the angular displacements of the driving mechanism θ_1 and the pinion θ_2 satisfy

$$I_1\ddot{\theta}_1 = K_1(\theta_2 - \theta_1). \tag{S1}$$

The pinion drives a gearwheel with an applied torque P where

$$I_2\ddot{\theta}_2 = K_1(\theta_1 - \theta_2) + P. \tag{S2}$$

If the angular displacement of the gearwheel is θ_3 and the gear ratio is γ, then for a perfect gear

$$\theta_2 + \gamma\theta_3 = 0 \tag{S3}$$

and the torque applied to the gearwheel is γP_3. This assumes a simple meshing between the pinion and gearwheel, with both having a large number of teeth which are always in contact, so that γ is the ratio of the mean radius of the gearwheel to that of the pinion.

If the output device has moment of inertia I_4, and is connected to the gearwheel of moment of inertia I_3 through a coupling modelled by a torsional stiffness K_2, then

$$I_3\ddot{\theta}_3 = K_2(\theta_4 - \theta_3) + \gamma P, \tag{S4}$$

and

$$I_4\ddot{\theta}_4 = K_2(\theta_3 - \theta_4), \tag{S5}$$

where θ_4 is the angular displacement of the output device. If P, K_1, and K_2 are eliminated between eqns (S1), (S2), (S4), and (S5) then

$$\gamma(I_1\theta_1 + I_2\theta_2) - I_3\theta_3 - I_4\theta_4 = kt, \tag{S6}$$

where k is a constant and the origin of time is suitably chosen. For a perfect gear a steady-motion solution of eqns (S1), (S3), (S5), and (S6), in which the angular velocities are all constant, is given by

$$\theta_1 = \theta_2 = -\gamma\Omega t, \quad \theta_3 = \theta_4 = \Omega t, \tag{S7}$$

where

$$k = -[\gamma^2(I_1 + I_2) + I_3 + I_4]\Omega.$$

The system of equations is linear, and we may look for periodic oscillations which can be superposed on this quasi-steady solution in the form $\bar{\theta}_j = a_j e^{i\omega t}$ so that

$$\begin{bmatrix} K_1 - \omega^2 I_1 & -K_1 & 0 & 0 \\ 0 & 1 & \gamma & 0 \\ 0 & 0 & -K_2 & K_2 - \omega^2 I_4 \\ \gamma I_1 & \gamma I_2 & -I_3 & -I_4 \end{bmatrix} \begin{bmatrix} a_1 \\ a_2 \\ a_3 \\ a_4 \end{bmatrix} = 0.$$

The condition that the determinant of this matrix must be zero determines the possible natural frequencies ω, and it is not difficult to show that there are two real values ω_1 and ω_2, given that all the constants are positive.

In practice the gearwheel does not perfectly mesh with the pinion and the perfect gear equation (S3) needs modification. There may be small irregularities in the teeth at manufacture, the gear-wheel axis may not be exactly parallel to that of the pinion, there may be significant elastic deformations of the teeth and long-term distortion, and there may be loss of contact between individual teeth for part of each cycle. To represent all these effects we consider a simple model of gearwheel wear in which the perfect gear equation (S3) is replaced by

$$\theta_2 + \gamma\theta_3 = \varepsilon f(\theta_3), \tag{S8}$$

where $f(\theta_3)$ is a periodic function, period 2π, and ε is a small parameter indicating the order of magnitude of the error. The wear $f(\theta_3)$ may be decomposed into Fourier components, and we may expect that the most significant contribution will arise from the first component. Hence, choosing the origin of θ_3 appropriately we shall consider only the wear term $\varepsilon \sin \theta_3$, and we shall look for an expansion in powers of ε. Since the solution with $\varepsilon = 0$ is $\theta_3 = \Omega t$, a regular perturbation will introduce a forcing term proportional to $\sin \Omega t$ which will create resonances with the natural frequencies ω_1 and ω_2. Thus for values of Ω not close to ω_1 or ω_2 the effect of the wear is $O(\varepsilon)$, and is not important. However when $\Omega \sim \omega_1$ or ω_2 there will be changes in the motion and torque larger than $O(\varepsilon)$. We therefore examine the motion for values of Ω near a resonance, and to do this we use the method of two-timing. First, however, the equations must be rewritten in a more suitable form; in so doing the problem can be generalized to more complicated systems, modelling gearboxes with more than one gearwheel and pinion.

If $\bar{\theta}_i$ are the perturbations from the quasi-steady state given by eqn (S7) then we must distinguish between external variables $\bar{\theta}_1$ and $\bar{\theta}_4$, whose second derivatives appear in the system of equations (S1), (S8), (S5) and (S6), and gearbox variables $\bar{\theta}_2$ and $\bar{\theta}_3$ which appear only as undifferentiated terms. Thus if x represents the external variables and y the gearbox variables, a more general problem is to solve

$$\ddot{x} + \varepsilon\mu\dot{x} + Ax + By = 0, \tag{S9}$$

subject to

$$Cx + Dy = \varepsilon f(y, t), \tag{S10}$$

where a small damping term has been introduced. In general x and y will be of even but different dimension, say n and m respectively, with f of dimension m; $A(n \times n)$, $B(n \times m)$, $C(m \times$

n), $D(m \times m)$ are appropriate real matrices and D is non-singular. In the simple model,

$$m = n = 2, \quad y = \begin{pmatrix} \bar{\theta}_2 \\ \bar{\theta}_3 \end{pmatrix}, \quad f = \begin{pmatrix} 0 \\ \sin(\Omega t + \bar{\theta}_3) \end{pmatrix}, \quad D = \begin{pmatrix} \gamma I_2 & -I_3 \\ 1 & \gamma \end{pmatrix},$$

which is clearly non-singular. With $\varepsilon = 0$, $y = D^{-1}Cx$ and

$$\ddot{x} + (A - BD^{-1}C)x = 0,$$

so that the natural frequencies are the zeros of

$$\det(A - BD^{-1}C - \omega^2 I),$$

and we assume that n distinct real zeros ω_i exist.

To examine a resonance near $\Omega = \omega_i$ it is convenient to take $\omega_i t$ as our time scale (thus modifying the entries in A and B by a factor ω_i^2), so that the determinant has a zero $\omega_i = 1$, and to consider $\Omega = 1 + \varepsilon k$. A second time scale $\tau = \varepsilon t$ is introduced so that $\mathrm{d}/\mathrm{d}t$ is replaced by $(\partial/\partial t + \varepsilon(\partial/\partial \tau))$, and eqn (S9) becomes

$$\frac{\partial^2 x}{\partial t^2} + 2\varepsilon \frac{\partial^2 x}{\partial t \, \partial \tau} + \varepsilon \mu \frac{\partial x}{\partial t} + Ax + By = \mathrm{O}(\varepsilon^2).$$

Expanding in powers of ε, the terms independent of ε satisfy

$$\frac{\partial^2 x_0}{\partial t^2} + Ax_0 + By_0 = 0 = Cx_0 + Dy_0.$$

Hence

$$x_0 = [a(\tau)\cos t + b(\tau)\sin t]e, \tag{S11}$$

where e is the unique unit eigenvector satisfying

$$(-I + A - BD^{-1}C)e = 0, \tag{S12}$$

since the eigenvalue is $+1$ and is distinct. From the terms of $\mathrm{O}(\varepsilon)$,

$$\frac{\partial^2 x_1}{\partial t^2} + Ax_1 + By_1 + 2\frac{\partial^2 x_0}{\partial t \, \partial \tau} + \mu \frac{\partial x_0}{\partial t} = 0,$$

$$Cx_1 + Dy_1 = f(y_0, t).$$

Eliminating y_1 (and y_0),

$$\frac{\partial^2 x_1}{\partial t^2} + (A - BD^{-1}C)x_1 = -BD^{-1}f(-D^{-1}Cx_0, t) - 2\frac{\partial^2 x_0}{\partial t \, \partial \tau} - \mu \frac{\partial x_0}{\partial t}.$$

To avoid secular terms in x_1 we require that the component of the right hand side in the direction of e does not contain terms in $\sin t$ or $\cos t$ when expanded as a Fourier series. This gives

$$\frac{\mathrm{d}a}{\mathrm{d}\tau}+\frac{\mu a}{2}=\frac{1}{2\pi}\int_{-\pi}^{\pi} BD^{-1}f \cdot e \sin t \, \mathrm{d}t,$$

$$\frac{\mathrm{d}b}{\mathrm{d}\tau}+\frac{\mu b}{2}=-\frac{1}{2\pi}\int_{-\pi}^{\pi} BD^{-1}f \cdot e \cos t \, \mathrm{d}t,$$

(S13)

where for convenience the limits have been taken from $-\pi$ to π. Whatever the size of the original system of external and gearbox variables, the perturbation problem reduces to solving a pair of first-order equations for the amplitudes a and b of the resonant mode, and depends solely on the scalar function $[BD^{-1}f] \cdot e$. The resonant mode of eqn (S11) is then an O(1) oscillation which can be sustained for all time if it can be initiated. Naturally we only expect to have initially a small perturbation of the steady-state motion, so we are interested in the solution curves of eqn (S13) in the phase-plane (a, b) which at $\tau = 0$ have small values of a and b, that is they are close to the origin of the phase plane.

For the simple model described earlier in this section with $n = m = 2$, $[BD^{-1}f] \cdot e$ has the form

$$c \sin[t + \kappa\tau + da(\tau)\cos t + db(\tau)\sin t],$$

(S14)

where c and d are given constants evaluated from A, B, C and D. The parameters κ and μ represent the effects of detuning and damping respectively, and we may expect that the largest perturbation (and the simplest analysis) occurs when $\mu = \kappa = 0$.

For fixed values of κ and μ we examine the phase plane in polar coordinates $a = (r/d)\sin\psi$, $b = (r/d)\cos\psi$ so that $x_0 = (r/d)\sin(t + \psi)e$. Then

$$\frac{\mathrm{d}r}{\mathrm{d}\tau}+\frac{\mu r}{2}=-\frac{cd}{2\pi}\int_{-\pi}^{\pi}\cos(t + \psi)\sin[t + \kappa\tau + r\sin(t + \psi)] \, \mathrm{d}t$$

and

$$r\frac{\mathrm{d}\psi}{\mathrm{d}\tau}=+\frac{cd}{2\pi}\int_{-\pi}^{\pi}\sin(t + \psi)\sin[t + \kappa\tau + r\sin(t + \psi)] \, \mathrm{d}t.$$

These integrals have the same value for any 2π range, so with limits $-\pi + \psi$ and $\pi + \psi$ and the change of variable $t + \psi = t'$ they reduce to

$$\int_{-\pi}^{\pi} \sin(t' + \kappa\tau - \psi + r\sin t') \begin{pmatrix} -\cos t' \\ \sin t' \end{pmatrix} dt'.$$

Expanding by the sine addition formulae, and noting that antisymmetric integrals are zero whereas symmetric ones double up, the integrals reduce to

$$-2\int_{0}^{\pi} \cos(t' + r\sin t')\sin(\kappa\tau - \psi)\cos t' \, dt'$$

and

$$2\int_{0}^{\pi} \sin(t' + r\sin t')\cos(\kappa\tau - \psi)\sin t' \, dt'.$$

These integrals are related to the Bessel function $J_1(r)$ defined by $(1/\pi)\int_0^{\pi}\cos(t' + r\sin t')\,dt$, so that finally

$$\frac{dr}{d\tau} + \frac{\mu r}{2} = -\frac{cd}{r}J_1(r)\sin(\psi - \kappa\tau),$$

$$r\frac{d\psi}{d\tau} = -cdJ_1'(r)\cos(\psi - \kappa\tau). \tag{S15}$$

With $\psi - \kappa\tau = \phi$, eqn (S15) reduces to a pair of autonomous first-order equations for r and ϕ, and there are many methods available for their solution. In the case $\mu = \kappa = 0$ they integrate simply to give

$$J_1(r)\cos\psi = \text{constant.} \tag{R16}$$

Integral curves are shown in Fig. S2, and those which pass close to the origin (in the (a, b) phase plane) necessarily lie for all τ within the circle $r = r_0$ where $J_1(r_0) = 0$, that is $r_0 = 3.83$.

It is thus possible for a small disturbance at a time $\tau = 0$ to produce an oscillation x_0 whose amplitude will in each period grow almost as large as $3.83/d$. Such an oscillation will give rise to $O(1)$ periodic changes in the torque P acting on the gearwheel, whose maximum value can be calculated, and this will be an important consideration in any design study.

With $\mu = 0$, $\kappa \neq 0$, integral curves are given by

$$J_1(r)\cos(\psi - \kappa\tau) - \frac{\kappa r^2}{2} = \text{constant,} \tag{S17}$$

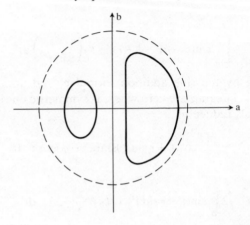

Fig. S2. Phase plane for eqn (S15), $\mu = \kappa = 0$

and the maximum amplitude is given by a root of $J_1(r) \pm \kappa r^2/2 = 0$. By carefully considering the phase plane it is possible to show that the maximum amplitude of x_0 decreases as $|\kappa|$ increases from zero; $\mu \neq 0$ may also be considered in a similar way, although in this case no closed-form integral exists.

A more interesting case is when Ω is continuously varied through its resonant value of unity. Thus if we assume that $\Omega = 1 + \varepsilon^2 \beta t$, then in eqns (S15) the detuning term $\kappa \tau$ is replaced by $\beta \tau^2$. Then with $\mu = 0$ and $\phi = \psi - \beta \tau^2$ the equations become

$$\frac{dr}{d\tau} = -\frac{cd}{r} J_1(r) \sin \phi,$$

$$\frac{d\phi}{d\tau} = \frac{-cd}{r} J_1'(r)\cos \phi + 2\beta\tau.$$

This pair of equations is not autonomous, but may still be tackled numerically by finite difference methods. There is now no longer any guarantee that the motion will be periodic in τ, and simple bounds on the amplitude cannot be found. Moreover it is possible that for some value of τ there will be a discontinuity in the dependent variables, due to the highly non-linear behaviour of the system.

5.2. Thin layers and lubrication theory

In previous chapters we have in several examples obtained simplified models by assuming that length scales in (say) the z direction are much smaller than those normal to it. The formal procedure used is to rescale the z variable with a small parameter ε expressing the ratio of the relative length scales. In aerodynamics this is used to develop the theory of flow past thin wings and the small parameter, which in this case is the ratio of the maximum thickness of the wing to its maximum chord length, is called an *aspect ratio*. This theory is well established (see for example Ockendon and Tayler (1983)), and the procedure may be used in many other applications.

In the problem of convective diffusion with a large Peclet number q, given by eqn (4.28), a rescaling with small parameter $1/q$ led to the simpler model of eqn (4.30). In Problem N, concerned with smoke dispersion from a chimney of height h in a wind of strength U, the solution was sought in a region where the downstream scale was much larger than the scale of the cross-flow, and the small parameter was Uh/D. In both cases an elliptic problem reduces to a easier parabolic problem, but downstream boundary conditions could no longer be posed.

In these three applications the domain is unbounded, and it is the disturbance which has length scales of different orders of magnitude. A more common use of the scaling procedure occurs when the domain itself has different length scales in different directions, that is the domain is a thin layer or a slender tube. In the simplest case for a thin layer the boundaries will be parallel planes and the solution will depend only on the variable z normal to the boundaries, as for example in the well-known Poiseuille flow of a viscous fluid. A more complicated version of this flow, when the viscosity varies exponentially with temperature, is discussed in Chapter 3 in the paragraphs leading up to eqns (3.34) and (3.35).

If the flow between the two planes is not unidirectional so that the solution is not independent of x or y, the problem may still be significantly simplified; with constant viscosity and slow flow, the Navier–Stokes equations reduce to

$$\frac{\partial^2 \bar{p}}{\partial x^2} + \frac{\partial^2 \bar{p}}{\partial y^2} = \mathrm{O}(\varepsilon^2),$$

220   *Asymptotic methods for non-linear problems*

as in eqns (3.9) and (3.10). The mean values $\hat{u}$ and $\hat{v}$ across the layer also satisfy Laplace's equation in the variables x and y parallel to the walls. This gives rise to a close analogy with the two-dimensional irrotational flow of an inviscid fluid, and is the basis of the Hele–Shaw apparatus for visualizing such a flow.

On a free boundary in a Hele–Shaw flow there will be, in addition to $\bar{p} = 0$, a kinematic condition given by eqn (3.11) in the form

$$\left(\frac{\partial\bar{p}}{\partial x}\right)^2 + \left(\frac{\partial\bar{p}}{\partial y}\right) = 12\frac{\partial\bar{p}}{\partial t}.$$

If the layer is thin but has variable width the procedure is still effective, and is commonly used in the modelling of the flow of oil in a bearing, where it is called *lubrication theory*. With the same scaling as for the Hele–Shaw flow, where now the small parameter is the ratio of the mean width of the layer to a typical lateral dimension, we obtain the same partial differential equations but with different boundary conditions on the walls. Without loss of generality we may take one of the walls as $\bar{z} = 0$, since we can think of $(x, y, \bar{z})$ as a curvilinear coordinate system as in Fig. 5.3 which has $\bar{z} = 0$ as one wall. In these curvilinear variables eqns (3.9) will be unaltered correct to $O(\varepsilon)$, provided that the boundary shape only changes on the scale of x and y. A variety of boundary conditions may be considered, but the simplest problem is to have a steady relative motion of the two boundaries given by $u = v = \bar{w} = 0$ on $\bar{z} = h$; $u = 1, v = 0 = \bar{w}$ on $\bar{z} = 0$. Then

$$u = \frac{\partial\bar{p}}{\partial x}\frac{\bar{z}^2 - h\bar{z}}{2} + 1 - \frac{\bar{z}}{h}, \tag{5.8}$$

$$v = \frac{\partial\bar{p}}{\partial y}\frac{\bar{z}^2 - h\bar{z}}{2}, \tag{5.9}$$

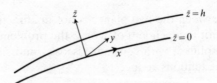

Fig. 5.3. Thin layer coordinates

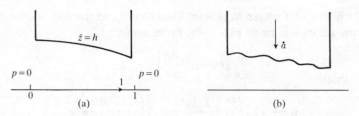

Fig. 5.4. Slider and squeeze film bearings

and

$$-\bar{w} = \int_0^{\bar{z}} \left[\left(\frac{\partial^2 \bar{p}}{\partial x^2} + \frac{\partial^2 \bar{p}}{\partial y^2} \right) \frac{(\bar{z}^2 - h\bar{z})}{2} - \left(\frac{\partial \bar{p}}{\partial x} \frac{\partial h}{\partial x} + \frac{\partial \bar{p}}{\partial y} \frac{\partial h}{\partial y} \right) \frac{\bar{z}}{2} + \frac{\bar{z}}{h^2} \frac{\partial h}{\partial x} \right] d\bar{z}.$$

Using the boundary condition $\bar{w} = 0$ on $\bar{z} = h$, $\bar{p}$ satisfies

$$\frac{\partial}{\partial x} \left(h^3 \frac{\partial \bar{p}}{\partial x} \right) + \frac{\partial}{\partial y} \left(h^3 \frac{\partial \bar{p}}{\partial y} \right) = 6 \frac{\partial h}{\partial x}. \tag{5.10}$$

This is called Reynolds equation, and it reduces to eqn (3.10) when h is constant. With a one-dimensional channel $\bar{z} = h(x)$ it represents the mass-conservation statement

$$\frac{d}{dx} \left(h - \frac{h^3}{6} \frac{d\bar{p}}{dx} \right) = 0, \tag{5.11}$$

which is easily integrated. Two boundary conditions for p are needed; for a simple slider-bearing as in Fig. 5.4a, the high pressures (p is $O(1/\varepsilon^2)$) must vanish at the ends of the bearing, so that $\bar{p} = 0$ at $x = 0$ and $x = 1$.

If the walls have a normal component of relative velocity of $O(\varepsilon)$, the problem will be unsteady but may still remain a slow flow; that is on the time scale L/U, where L is the length of the bearing time derivatives will appear only in the boundary conditions; and on $\bar{z} = h$, $\bar{w} = \partial h/\partial t$ and $u = v = O(\varepsilon)$. The Reynolds equation (in one space-variable) now becomes

$$\frac{1}{6} \frac{\partial}{\partial x} \left(h^3 \frac{\partial \bar{p}}{\partial x} \right) = \frac{\partial h}{\partial x} + 2 \frac{\partial h}{\partial t}. \tag{5.12}$$

A *squeeze film* occurs when a rigid block $0 < x < 1$ moves towards a fixed wall $\bar{z} = 0$. In such a situation the surface of the block, assumed to be one-dimensional for simplicity, is given by

$\bar{z} = h(x) - a(t)$ where $h(x)$ is its fixed shape and the gap is closing with velocity $\dot{a}$ as in Fig. 5.4b. From eqn (5.12),

$$\frac{d}{dx}\left[(h-a)^3\frac{d\bar{p}}{dx}\right] = -12\dot{a}$$

so that

$$\bar{p} = -12\dot{a}\int_0^x \frac{x\,dx}{[h(x)-a]^3} + A\int_0^x \frac{dx}{[h(x)-a]^3},$$

where A is determined by the conditions $\bar{p} = 0$ at $x = 1$. The total force on the block will be given in non-dimensional form by $F(t) = \int_0^1 \bar{p}\,dx$, since in the lubrication approximation the stress tensor is dominated by the pressure term. If the block has mass M and is supported by the squeeze film, then

$$\int_0^1 \bar{p}\,dx = \frac{Mg\varepsilon^2}{\mu U} = \lambda,$$

where λ is an $O(1)$ non-dimensional parameter. Here U is a typical velocity scale in the x direction, and since the typical velocity scale in the $\bar{z}$ direction is $(g\varepsilon L)^{\frac{1}{2}}$ we choose $U = (gL/\varepsilon)^{\frac{1}{2}}$ so that

$$\lambda = \frac{M}{\mu}\left(\frac{\varepsilon^5 g}{L}\right)^{\frac{1}{2}},$$

implying that $\mu^2 L/M^2 g \ll 1$ for this squeeze-film motion to be possible. On substituting for p we obtain an integro-differential equation for the block settling motion $a(t)$, where at $t = 0$ we choose $a = 0$. In the case of a plane block parallel to $\bar{z} = 0$ so that $h = 1$, $\bar{p} = 6\dot{a}(x - x^2)/(1-a)^3$ and $\dot{a}/(1-a)^3 = \lambda$. Hence $(1-a) = (1 + 2\lambda t)^{-\frac{1}{2}}$ and the gap takes an infinite time to close.

A related problem, in which the plane block is pivoted at its mid-point and the wall $\bar{z} = 0$ moves with unit speed, is given in Exercise 5.2a. A variety of bearing problems are discussed by Cameron (1981), all of which lead to a Reynolds-type equation. The gas bearing for which the fluid density is proportional to the pressure is another interesting example, and leads to the non-linear equation

$$\frac{1}{6}\frac{d}{dx}\left(h^3\bar{p}\frac{d\bar{p}}{dx}\right) = \bar{p}\frac{dh}{dx} - h\frac{d\bar{p}}{dx}, \qquad (5.13)$$

obtained from the compressible form of the continuity equation.

Thin layers may occur with significantly different boundary conditions to those encountered in modelling bearings. The scalings required will be different, but in many cases still lead to an equation of Reynolds type. One example occurred in Problem R, where the thin layer was an oil film and one wall was replaced by a free boundary on which $p = 0$. Clearly in this case p is at most O(1) in the fluid, so that the non-dimensional velocities u and v must be small. They should be re-scaled by $\bar{u} = u/\varepsilon^2$, $\bar{v} = v/\varepsilon^2$, leaving p unscaled to give equations identical with eqn (5.7) in terms of $\bar{u}$, $\bar{v}$ and p, except that in this problem the driving mechanism is the hydrostatic pressure and

$$\frac{\partial p}{\partial \bar{z}} = -\frac{\rho g \varepsilon L^2}{\mu U} = -\bar{G}.$$

$\bar{G}$ has to be an O(1) non-dimensional parameter, and since a typical velocity in the $\bar{z}$ direction is $(g\varepsilon L)^{\frac{1}{2}}$ we again choose $U = (gL/\varepsilon)^{\frac{1}{2}}$ to give $\bar{G} = (\rho/\mu)(g\varepsilon^3 L^3)^{\frac{1}{2}}$, implying that $\mu^2/g\rho^2 L^3 \ll 1$ for the thin film to exist. Note that there is no unique way of making u, v and p non-dimensional, and that in Problem R a different choice was made leading to a different parameter G equal to $\bar{G}^2$; but both are O(1), so the equations and conclusions are unchanged. The Reynolds equation has the form of eqn (R4) and is a non-linear equation for the layer thickness, since the pressure on the layer is prescribed.

Another example is a shear layer; this occurred in Problem Q, when the molten flow upstream of the laser was considered in the case of large Peclet number. In this problem $\bar{z} = 0$ moves with an O(1) velocity which is not tangential to the wall, and w is O(1) in the layer. From continuity u and v will be O(1/ε) and we rescale by $\bar{u} = \varepsilon u$, $\bar{v} = \varepsilon v$; we now choose $\bar{p} = \varepsilon^3 p$ to give eqns (5.7) again, but in terms of $\bar{u}$, $\bar{v}$, w and $\bar{p}$. The boundary conditions are $\bar{u} = \bar{v} = w = 0$ on $\bar{z} = h$; $\bar{u} = \bar{v} = 0$, $w = Q_n$, (the normal component of the given velocity) on $\bar{z} = 0$. Note that the tangential components of $\mathbf{Q}$ do not appear in the boundary conditions for $\bar{u}$ and $\bar{v}$ to this order of approximation, and the pressure is much larger than in the relative tangential motion case. The Reynolds equation is easily derived in the form

$$\frac{\partial}{\partial x}\left(h^3 \frac{\partial \bar{p}}{\partial x}\right) + \frac{\partial}{\partial y}\left(h^3 \frac{\partial \bar{p}}{\partial y}\right) = -6Q_n \qquad (5.14)$$

In Problem Q the layer thickness is, however, unknown and is to be determined from the lubrication form of the energy equation and a Stefan condition on $y = h$. A further example of such a shear layer is given in Exercise 5.2b.

The scaling will also be different in a channel which has parallel walls for some value of x, and the most obvious example is a flow between a pair of rollers of large radius, or with symmetry the flow between a roller and a plane. This has been briefly discussed in Problem M for the hot rolling of a steel ingot, but we re-examine the situation in the context of the slow-motion equations. If the region of interest is that under the rollers near the nip (the minimum gap) as in Fig. M1, then the slope will be small there and we may expect that the induced pressure may not be as large as it would be in the usual lubrication theory of bearings. If we take as the smaller parameter the ratio of the minimum gap (between the roller and $\bar{z} = 0$) to the roller radius, then for the circular roller

$$[1 + \varepsilon(1 - h)]^2 + x^2 = 1,$$

where x has been scaled with the radius R. Expanding this for small ε, $h = 1 + x^2/2\varepsilon + O(\varepsilon)$, so that we must rescale x by $\bar{x} = x/\varepsilon^{\frac{1}{2}}$ to give a channel $h = 1 + \bar{x}^2/2$. The velocity u will be made non-dimensional with ΩR, where Ω is the angular velocity of the roller, and is clearly $O(1)$. Hence to obtain eqns (5.7) in terms of $\bar{x}$ the appropriate scaling for $\bar{p}$ is $p\varepsilon^{\frac{3}{2}}$. If both the roller surface and the channel centre-line have the same speed, the boundary conditions are $\bar{z} = 0$, $u = 1$, $\bar{w} = 0$; $\bar{z} = 1 + \bar{x}^2/2$, $u = 1$, $\bar{w} = \bar{x}$. The appropriate Reynolds equation is

$$\frac{\mathrm{d}}{\mathrm{d}\bar{x}} \left(h - \frac{h^3}{12} \frac{\mathrm{d}\bar{p}}{\mathrm{d}\bar{x}} \right) = 0, \tag{5.15}$$

which has a factor of 2 different from eqn (5.11) because in this problem both boundaries are moving. The obvious boundary conditions on $\bar{p}$ are that $\bar{p} = 0$ at the ends of the roller channel, that is as $\bar{x} \to \pm\infty$. A typical profile for $\bar{p}$ is shown in Fig. 5.5a; it is antisymmetric about $\bar{x} = 0$. This implies that the actual pressure (p was made non-dimensional in terms of the pressure above ambient) will be negative for most of the region downstream of the nip, which is not physically acceptable. A cavity must form as in Fig. 5.5b at some point $\bar{x} = d$ downstream of the nip.

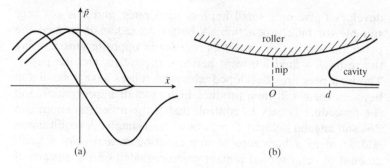

Fig. 5.5. Flow under a roller

Clearly $\bar{p} = 0$ at $\bar{x} = d$, but to determine the solution a further condition is needed. Detailed studies of bubbles in narrow channels indicate that the pressure gradient $d\bar{p}/d\bar{x}$ at the tip $\bar{x} = d$ is a non-negative function of the ratio of $\frac{1}{2}h(d)$ to the half layer thickness as $\bar{x} \to \infty$. This implies a small region where $\bar{p} < 0$ as in Fig. 5.5a, but the actual pressure may still be positive if the ambient pressure is large enough. However a simpler condition, which is asymptotically consistent with this cavity condition, is to put $d\bar{p}/d\bar{x} = 0$ at $\bar{x} = d$ with $\bar{p} > 0$ for $\bar{x} < d$. Then from eqn (5.15)

$$\left(1 + \frac{\bar{x}^2}{2}\right)^3 \frac{d\bar{p}}{d\bar{x}} = 6(\bar{x}^2 - d^2),$$

and d is determined from

$$\int_{-\infty}^{d} \frac{(\bar{x}^2 - d^2)}{(2 + \bar{x}^2)^3} \, d\bar{x} = 0$$

to complete the solution.

Problem T. *The impregnation of paper by a resin*

In manufacturing laminates it is required to uniformly impregnate sheets of absorbent paper with a given amount per unit area of a resin. The sheets are then pressed together to form the final material. One simple procedure for impregnating the paper is to feed the paper into the bottom of a tank of the resin, and use the effects of capillarity and hydrostatic pressure to achieve impregnation. This is, however, a slow procedure since the pressures developed give only small impregnation rates, and it is not very suitable for bulk production. A better procedure is to feed the paper through a pair of rollers rotating in opposite senses, using the resin as a fluid lubricant between the rollers and the paper. The high pressures developed when the rollers have a small gap between them will then produce high rates of impregnation and the procedure is easy to control, that is the roller minimum gap $2h_0$ and angular velocity Ω may be altered simply. A requirement additional to a high rate of impregnation is that only a small amount of resin should remain unimpregnated on the surface of the paper, to improve the drying process and avoid wastage. Thus a model of this process should attempt to predict m, the fluid mass impregnated per unit area, and M, the residual mass of fluid per unit area on the surface, in terms of h_0 and Ω and the physical properties of the paper and fluid. As in the case of classical lubrication theory we expect a cavity to form in the downstream side of the roller, and the configuration is shown in Fig. T1. The paper will deform under the high pressures developed in the resin, and in the steady state will have a non-dimensional thickness in lubrication-theory scaled variables of $(s_0 - s)$, so that the dimensional unperturbed paper thickness is $2s_0 h_0$. A simple model for the compressibility of the paper is to assume a linear relation between the high positive pressure at the surface and the surface displacement. Thus we define a compressibility coefficient c by the relation

$$s = cp, \tag{T1}$$

where p is the scaled non-dimensional lubrication pressure (corresponding to $\bar{p}$ in eqn (5.15)). The coefficient c is given by $\mu\Omega R^{\frac{3}{2}}C/h_0^{\frac{5}{2}}$, where C is a dimensional elastic constant relating the dimensional pressure above ambient to the actual displacement,

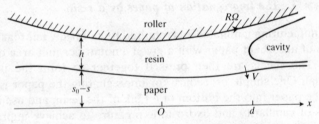

Fig. T1. Rolling a porous compressible paper

and is a given property of the paper. Because of the variation in the paper thickness due to the changing pressure, the fluid channel thickness h will vary and will be given by

$$h = 1 - s_0 + \frac{x^2}{2} + cp. \tag{T2}$$

Since the paper is porous, resin flows into it at a non-dimensional rate V which depends on the surface pressure. A simple model is to assume a linear relation $V = kp$ where k is a porosity coefficient given by RK/h_0^2 and K is a dimensional constant relating the dimensional pressure above ambient to the actual impregnation rate, which is a property of the paper and resin determined from static tests. Thus the fluid channel loses mass flow at a rate V and Reynold's equation (5.15), which is a mass-flow balance, becomes

$$\frac{d}{dx} \left(h - \frac{h^3}{12} \frac{dp}{dx} \right) = -V. \tag{T3}$$

Eliminating h we obtain

$$\frac{1}{12} \frac{d}{dx} \left[\left(1 - s_0 + \frac{x^2}{2} + cp \right)^3 \frac{dp}{dx} \right] = kp + x + c\frac{dp}{dx}, \tag{T4}$$

together with the boundary conditions as before which are

$$p(-\infty) = p(d) = \frac{dp}{dx}(d) = 0. \tag{T5}$$

This is a non-linear second-order equation for p for which the coefficient of the highest derivative d^2p/dx^2 could vanish, implying that the channel closes and the paper touches the roller. Such an operating condition would be highly undesirable since the paper would stick to the roller and tear, and a further necessary constraint is that $1 - s_0 + x^2/2 + cp > 0$.

However, since the pressure increases as the fluid channel narrows it is likely that the optimum operating conditions will occur when $(1 - s_0 + x^2/2 + cp)$ is small, so that eqn (T4) is a singular-perturbation problem for which standard numerical procedures for second-order equations will run into difficulties. We therefore attempt a perturbation solution when the compressibility and porosity coefficients c and k are both small (which in practice is usually the case), and write $\delta = s_0 - 1 > 0$ so that the uncompressed paper is thicker than the gap between the rollers.

Equation (T4) may then be reduced to a problem independent of δ by the transformation $x = \delta^{\frac{1}{2}}\bar{x}$, $\bar{p} = \delta^{\frac{3}{2}}p$ so that

$$\frac{1}{12}\frac{d}{d\bar{x}}\left[\left(-1+\frac{\bar{x}^2}{2}+\bar{c}\bar{p}\right)^3\frac{d\bar{p}}{d\bar{x}}\right] = \bar{k}\bar{p}+\bar{x}+\bar{c}\frac{d\bar{p}}{d\bar{x}}, \tag{T6}$$

with boundary conditions $\bar{p}(-\infty) = 0 = \bar{p}(\bar{d}) = \dfrac{d\bar{p}}{d\bar{x}}(\bar{d})$, where

$$k = \delta^2\bar{k}, \quad c = \delta^{\frac{3}{2}}\bar{c}. \tag{T7}$$

A regular expansion in $\bar{c}$ and $\bar{k}$ will have a zero-order term satisfying

$$\left(-1+\frac{\bar{x}^2}{2}\right)^3\frac{d\bar{p}}{d\bar{x}} = 6\bar{x}^2 + \text{constant},$$

which will only be valid for $\bar{x} < -\sqrt{2}$, or possibly $\bar{x} > \sqrt{2}$ if $\bar{d} > \sqrt{2}$. We have therefore to consider boundary-layer (or rather internal-layer) regions near $\bar{x} = \pm\sqrt{2}$, and the region $-\sqrt{2} < \bar{x} < \sqrt{2}$ joining them; we then attempt to match all four or five regions together as in Fig. T2.

The difficulty in carrying out this matched expansion problem is twofold. Firstly there are several different regions, not just two, and the relative scales of p in each region are not obvious. More difficult, however, is the fact that there are two small parameters, and for different relative sizes of these parameters there will be

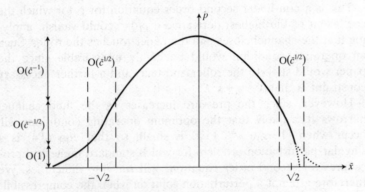

Fig. T2. Solution domains for eqn (T6)

different forms of expansion. We are interested in conditions close to the critical when the channel just closes, since in such conditions almost all the resin will be impregnated. Increasing the porosity and decreasing the compressibility both tend to close the channel, so that there will be a critical balance when $\bar{k} = \lambda \bar{c}^m$ where λ and m have to be determined. With considerable hindsight we choose $m = \frac{3}{2}$ to avoid the difficulty of two small parameters, and focus on the details of the matched expansions. Without this hindsight the various possibilities for any value of m must each be explored, and it may be demonstrated that only $m = \frac{3}{2}$ achieves the required critical conditions. We shall therefore be content with demonstrating one solution which is uniformly valid, and gives a critical value for λ above which the channel closes. We examine each region in turn.

(i) $-\infty < \bar{x} < -\sqrt{2}$. In this region

$$\left(-1 + \frac{\bar{x}^2}{2}\right)^3 \frac{d\bar{p}}{d\bar{x}} = 12\left(-1 + \frac{\bar{x}^2}{2}\right) - 4a\sqrt{2}. \tag{T8}$$

Here a is a constant to be determined by the matching as $\bar{x} \to -\sqrt{2}$, where after some simple manipulation

$$\bar{p} \sim \frac{-6}{\bar{x} + \sqrt{2}} - \frac{a}{(\bar{x} + \sqrt{2})^2}. \tag{T9}$$

(ii) $\bar{x} \sim -\sqrt{2}$. To avoid the coefficient of $d^2\bar{p}/d\bar{x}^2$ becoming zero, $\bar{c}\bar{p}$ must be of the same order of magnitude as $(-1 + \bar{x}^2/2)$. With $\bar{x} + \sqrt{2} = X\bar{c}^n$ and $\bar{c}^{1-n}\bar{p} = f + 2^{\frac{1}{2}}X$, this is satisfied and is a thin layer if $n > 0$. Then

$$\bar{c}^{2n-1} \frac{d}{dX}\left\{[f + O(\bar{c}^n)]^3 \frac{d}{dX}\right\}(f + 2^{\frac{1}{2}}X) = 12\frac{df}{dX} + O(\bar{c}^n).$$

To retain a balance of at least two terms we choose $n = \frac{1}{2}$ so that

$$f^3 \frac{df}{dX} = 12f - 2^{\frac{1}{2}}f^3 - 12b\sqrt{2}, \tag{T10}$$

where b is another constant.

Matching conditions are to be applied as $X \to \pm\infty$ and the integral curves of (T10) are shown in the various cases in Fig. T3. There is one integral curve which is continuous from $X \to -\infty$ to

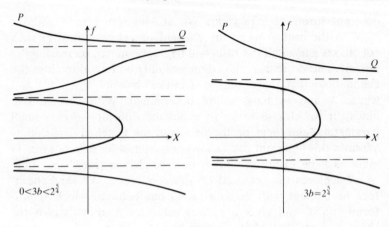

Fig. T3. Integral curves of eqn (T10)

$X \to +\infty$ and has $f + 2^{\frac{1}{2}}X$ positive for all X if $0 < 3b \le 2^{\frac{5}{4}}$, namely PQ. Thus as $X \to +\infty$, $f \to \beta$, where β is the largest root of the cubic

$$\beta^3 - (6\sqrt{2})\beta + 12b = 0, \tag{T11}$$

and is the matching condition with the region $\bar{x} > -\sqrt{2}$. As $X \to -\infty$, $f + 2^{\frac{1}{2}}X \to 0$ and it is easily shown that

$$f + 2^{\frac{1}{2}}X \sim -\frac{6}{X} - \frac{3b}{X^2}.$$

In the variables of region (i) this becomes

$$\bar{p} \sim -\frac{6}{\bar{x} + \sqrt{2}} - \frac{3b\bar{c}^{\frac{1}{2}}}{(\bar{x} + \sqrt{2})^2}$$

and the matching is achieved correct to the zero order by choosing $a = 0$ in eqn (T9). In this region near $\bar{x} = -\sqrt{2}$ the pressure is $O(\bar{c}^{-\frac{1}{2}})$ and grows even larger as $X \to +\infty$, that is in the region $-\sqrt{2} < \bar{x} < \sqrt{2}$.

(iii) $-\sqrt{2} < \bar{x} < \sqrt{2}$. In this region the channel will not close if $\bar{c}\bar{p} \sim 1 - \frac{1}{2}\bar{x}^2$, and there can be no scaling of the $\bar{x}$ variable. Hence we write $\bar{c}\bar{p} = 1 - \frac{1}{2}\bar{x}^2 + \bar{c}^{\frac{1}{2}}q$, so that the matching condition as $\bar{x} \to -\sqrt{2}$ can be satisfied by q. In the variables of region (ii), $q \sim f(\infty) = \beta$ so that $q(-\sqrt{2}) = \beta$ is the boundary condition for the equation

$$\frac{1}{12}\frac{d}{d\bar{x}}\{q^3[-\bar{x}+O(\bar{c}^{\frac{1}{2}})]\}=\lambda\left[1-\frac{\bar{x}^2}{2}+O(\bar{c}^{\frac{1}{2}})\right]+\frac{dq}{d\bar{x}},$$

which is obtained by substituting for q in eqn (T6). It is now clear why the relationship $\bar{k}=\lambda\bar{c}^{\frac{3}{2}}$ was critical. For smaller values of m there would have been a singular term on the right hand side as $\bar{c}\to 0$; for larger values the porosity term would not appear in this equation, or in the problem at all, to the zero order.

Integrating and using the condition at $\bar{x}=-\sqrt{2}$,

$$-\bar{x}q^3=2\lambda(6\bar{x}-\bar{x}^3)+12q+8\lambda\sqrt{2}-12b\sqrt{2}, \tag{T12}$$

so that as $\bar{x}\to+\sqrt{2}$, $q\to\gamma$ where

$$\gamma^3+(6\sqrt{2})\gamma-12b+16\lambda=0, \tag{T13}$$

and for a positive root to exist $3b>4\lambda$.

(iv) $\bar{x}\sim\sqrt{2}$. Here there is a boundary layer similar to that in region (ii), and we define $Y\bar{c}^{\frac{1}{2}}=\bar{x}-\sqrt{2}$, $\bar{c}^{\frac{1}{2}}\bar{p}=g-Y\sqrt{2}$ and we consider whether the cavity might begin to form in this region, that is look for $D\bar{c}^{\frac{1}{2}}=\bar{d}-\sqrt{2}$. The boundary conditions on g will then be $g(-\infty)=\gamma$, $g(D)=D\sqrt{2}$, $dg/dY=\sqrt{2}$. Substituting for g in eqn (T6) and integrating once we obtain

$$g^3\frac{dg}{dY}=12g+g^3\sqrt{2}-(12\sqrt{2})D. \tag{T14}$$

As $Y\to-\infty$, the integral curve of eqn (T14) which is continuous from $Y\to-\infty$ to $Y=D$ is such that $g\to\gamma$ where γ is the root of the cubic

$$\gamma^3+(6\sqrt{2})\gamma-12D=0,$$

so that $D>0$ because $\gamma>0$. By comparison with eqn (T13), $b=D+\frac{4}{3}\lambda>0$ and the zero-order solution is determined in all four regions in terms of the one unknown parameter D.

To complete the solution and determine D it is necessary to examine eqn (T12) in more detail, and it is sketched in Fig. T4. We require that q is a single-valued function which exists at all points $-\sqrt{2}<\bar{x}<\sqrt{2}$, that is the branch XY in Fig. T4. There will therefore in general be three roots for q at $x=-\sqrt{2}$, so that in

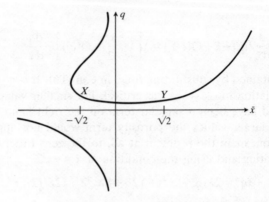

Fig. T4. Integral curves of eqn (T14)

Fig. T3 $3b < 2^{\frac{5}{4}}$, and the root which lies on XY is the one of intermediary value. But in determining the boundary layer near $\bar{x} = -\sqrt{2}$, the solution which was an integral curve of eqn (T10) (and which is sketched in Fig. T3) had a limit as $X \to \infty$ which was the largest root of the cubic equation (T11). The only way in which this discrepancy in the matching can be avoided is for there to be a double root of eqn (T11), that is for $3b = 2^{\frac{5}{4}}$.

This necessary condition determines D from

$$3D = 2^{\frac{5}{4}} - 4\lambda, \tag{T15}$$

which is valid provided $\lambda < 2^{-\frac{3}{4}}$, or $\bar{k} < (\bar{c}/\sqrt{2})^{\frac{3}{2}}$. It is a simple matter to calculate the mass-flow impregnated, since the dominant contribution comes from region (iii) where the pressure is greatest. The non-dimensional mass-flow impregnated is

$$\bar{m} = k \int_{-\infty}^{d} p \, \mathrm{d}x = \frac{k}{\delta} \int_{-\infty}^{d} \bar{p} \, \mathrm{d}\bar{x} = \left(\frac{4\sqrt{2}}{3}\right) \frac{\delta^{\frac{3}{2}} k}{c} + \text{smaller terms.}$$

The non-dimensional mass-flow residual at $x = d$, half of which may be assumed to stick to the surface of the paper, is

$$\bar{M} = \tfrac{1}{2}\delta(-1 + \tfrac{1}{2}\bar{d}^2) = \frac{c^{\frac{1}{2}}}{3}\left[\left(\frac{8}{\delta}\right)^{\frac{1}{4}} - k\left(\frac{\delta\sqrt{2}}{c}\right)^{\frac{3}{2}}\right].$$

In terms of dimensional quantities m and M

$$m = \left(\frac{4\sqrt{2}}{3}\right) \frac{K}{C} \frac{R^{\frac{1}{2}} h_0^{\frac{3}{2}}}{\Omega} (s_0 - 1)^{\frac{3}{2}},$$

$$M = \frac{2^{3/4} C^{1/2}}{3} \frac{(\mu \Omega)^{1/2} R^{3/4}}{h_0^{1/4} (s_0 - 1)^{1/4}} \left(1 - \frac{K 2^{3/4}}{C^{3/2}} \frac{h_0^{7/4} (s_0 - 1)^{7/4}}{\mu^{1/2} \Omega^{3/2} R^{5/4}}\right),$$

and the effects of varying h_0 and Ω may be evaluated.

Other regimes and improved models for the elasticity and porosity of the paper may also be considered; see Exercise 5.2c.

5.3. Boundary layers

For eqn (5.5) and Problem T, regular expansions have failed for particular values x_0 of the independent variable x. In both cases a new inner variable was defined which magnified the region near $x = x_0$, and an inner solution constructed which was matched to the regular outer solution. This singular situation may occur for problems in more than one variable, so that the singular region is in the neighbourhood of some curve if there are two independent variables. In many cases the singular region occurs because the solution has large derivatives near part of the boundary of the domain, and cannot be adequately represented by a regular outer expansion. The inner region will therefore be a thin layer, of varying thickness, adjacent to the boundary, and called a *bound-ary layer*. The appropriate stretched inner variable $\bar{n}$ will clearly be normal to the boundary and the matching will be between the inner solution as $\bar{n} \to \infty$ and the outer solution evaluated close to the boundary as $n \to 0$, as in Fig. 5.6.

Boundary layers occur in a wide range of models of physical problems, and are of particular importance in Fluid Mechanics for fluids of small viscosity, when in dimensionless variables the inverse of the Reynolds number is a small parameter.

A simple example is the impulsive motion U of an infinite flat plate adjacent to a half-space of slightly viscous fluid, the so-called Rayleigh problem. The flow may be described by $q = [u(y), 0, 0]$ where $y > 0$ is the fluid region and u satisfies

$$\frac{\partial u}{\partial t} = -\frac{\partial p}{\partial x} + \frac{1}{Re} \frac{\partial^2 u}{\partial y^2},$$

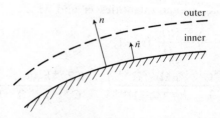

Fig. 5.6. Boundary layer coordinates

from the Navier–Stokes equations (1.2) in non-dimensional form (with infinite Froude number). Also $\partial p/\partial y = \partial p/\partial z = 0$, so that $\partial p/\partial x$ is a function only of time and, assuming that constant pressure is applied as $x \to \pm\infty$, may be put equal to zero.

The boundary conditions are $u = 1$, $y = 0$, $t > 0$; $u = 0$, $t = 0$, $y > 0$; $u = 0$, $y \to \infty$. A regular expansion in powers of $1/Re$ gives $u \equiv 0$, so that $\partial^2 u/\partial y^2$ must be large in some region which we expect to be adjacent to $y = 0$. Hence we rescale by writing $\bar{y} = (Re)^{\frac{1}{2}}y$ to obtain the inner problem

$$\frac{\partial u}{\partial t} = \frac{\partial^2 u}{\partial \bar{y}^2}; \quad \bar{y} > 0;$$

$$u = 1, \bar{y} = 0; \quad u = 0, t = 0;$$

and a matching condition that $u \to 0$ as $\bar{y} \to \infty$. This of course is the full problem, and in this very simple example the perturbation procedure has not simplified the problem as it would in a less simple example. However, we can solve this boundary-value problem and have done so already for eqn (4.30) using a similarity variable $\bar{y}/2\sqrt{t}$ to obtain, as in eqn (4.31),

$$u = 1 - \operatorname{erf}\left[\frac{y}{2}\left(\frac{Re}{t}\right)^{\frac{1}{2}}\right]. \tag{5.16}$$

Thus for $Re \gg 1$, u is only significantly different from zero (that is, more than exponentially small) in a region in which $y \sim O[(t/Re)^{\frac{1}{2}}]$, giving a boundary layer of thickness $O[(t/Re)^{\frac{1}{2}}]$ adjacent to $y = 0$. If the fluid were bounded by a second fixed parallel plate distant L from the moving plate as in Fig. 5.7a, the solution of eqn (5.16) would be asymptotically correct for large Reynolds number until the boundary-layer thickness was of comparable

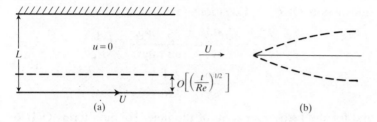

Fig. 5.7. Viscous flow past a semi-infinite flat plate

size to L, that is for non-dimensional times $t \ll Re$; or in dimensional variables, time much less than L^2/ν. For large enough times the Couette flow $u = 1 - y$ will be established.

A less straightforward but more instructive and important example is the steady flow U of a slightly viscous fluid past a semi-infinite flat plate, as in Fig. 5.7b. The flow is now two-dimensional, and in terms of a non-dimensional stream function ψ the steady Navier–Stokes equations may be written

$$\frac{\partial\psi}{\partial y}\frac{\partial^2\psi}{\partial x\,\partial y} - \frac{\partial\psi}{\partial x}\frac{\partial^2\psi}{\partial y^2} = -\frac{\partial p}{\partial x} + \frac{1}{Re}\left(\frac{\partial^2\psi}{\partial x^2\,\partial y} + \frac{\partial^3\psi}{\partial y^3}\right),$$

$$-\frac{\partial\psi}{\partial y}\frac{\partial^2\psi}{\partial x^2} + \frac{\partial\psi}{\partial x}\frac{\partial^2\psi}{\partial x\,\partial y} = -\frac{\partial p}{\partial y} + \frac{1}{Re}\left(-\frac{\partial^3\psi}{\partial x^3} - \frac{\partial^3\psi}{\partial x\,\partial y^2}\right),$$

(5.17)

with boundary conditions

$$\frac{\partial\psi}{\partial x} = \frac{\partial\psi}{\partial y} = 0 \quad \text{on} \quad y = 0, \quad x > 0;$$

$$\frac{\partial\psi}{\partial y} \to 1, \quad \frac{\partial\psi}{\partial x} \to 0 \quad \text{as} \quad y \to \pm\infty \quad \text{or} \quad x \to -\infty.$$

The outer solution with $1/Re = 0$ is clearly a uniform stream with $\psi = y$, but this does not satisfy the condition $\partial\psi/\partial y = 0$ on $y = 0$, $x > 0$. Hence we look for a singular region adjacent to $y = 0$, $x > 0$ with scaled variables $\bar{y}$, $\bar{\psi}$ and by symmetry need only consider $y > 0$. The matching requires that as $\bar{y} \to \infty$, $\partial\bar{\psi}/\partial\bar{y} \to 1$ so that the scaling for $\bar{\psi}$ and $\bar{y}$ must be the same. For a viscous and inertia term in the first equation of (5.17) to balance after such a scaling, the only choice is $\bar{y} = (Re)^{\frac{1}{2}}y$ and $\bar{\psi} = (Re)^{\frac{1}{2}}\psi$, so that as in the impulsive flat-plate problem the boundary-layer thickness is $O(Re^{-\frac{1}{2}})$. Equations (5.17) now reduce to

$$\frac{\partial \bar{\psi}}{\partial \bar{y}}\frac{\partial^2 \bar{\psi}}{\partial x\, \partial \bar{y}} - \frac{\partial \bar{\psi}}{\partial x}\frac{\partial^2 \bar{\psi}}{\partial \bar{y}^2} = -\frac{\partial p}{\partial x} + \frac{\partial^3 \bar{\psi}}{\partial \bar{y}^3} + O\!\left(\frac{1}{Re}\right),$$

$$\frac{\partial p}{\partial \bar{y}} = O\!\left(\frac{1}{Re}\right),$$

(5.18)

and for the zero-order term of the inner solution terms $O(1/Re)$ are omitted. The pressure is therefore constant across the boundary layer, and must match with the pressure in the outer region which is constant for all x because it is a uniform stream. Hence $\partial p/\partial x = 0$ everywhere in the layer, and we obtain the boundary-layer equation

$$\frac{\partial \bar{\psi}}{\partial \bar{y}}\frac{\partial^2 \bar{\psi}}{\partial x\, \partial \bar{y}} - \frac{\partial \bar{\psi}}{\partial x}\frac{\partial^2 \bar{\psi}}{\partial \bar{y}^2} = \frac{\partial^3 \bar{\psi}}{\partial \bar{y}^3},$$

(5.19)

with boundary conditions $\bar{\psi} = \partial\bar{\psi}/\partial\bar{y} = 0$ on $\bar{y} = 0$, $x > 0$, and matching condition $\partial\bar{\psi}/\partial\bar{y} \to 1$ as $\bar{y} \to \infty$.

This inner problem is a non-linear partial differential equation, difficult to solve, but nevertheless much simpler than the full problem of eqn (5.17). For these boundary conditions it admits a similarity solution, and we define a similarity variable $\eta = \bar{y}/x^{\frac{1}{2}}$. To satisfy the condition as $\bar{y} \to \infty$ with a similarity-type solution we require that $\partial\bar{\psi}/\partial\bar{y}$ is a function of η alone, and we write $\bar{\psi} = x^{\frac{1}{2}}f(\eta)$ so that $\partial\bar{\psi}/\partial\bar{y} = f'(\eta)$. After some manipulation eqn (5.19) reduces to

$$2f''' + ff'' = 0,$$

(5.20)

with boundary conditions $f(0) = f'(0) = 0$, $f'(\infty) = 1$; and the factor 2 in eqn (5.20) may be removed by modifying the definition of f and η to give Blasius' equation. The equation is autonomous and may be integrated once (see Exercise 5.3a); a numerical solution is also easily obtained. This boundary-layer solution is valid where η is $O(1)$, that is y is $O[(x/Re)^{\frac{1}{2}}]$ so that it grows parabolically with x just as in the Rayleigh solution it grew parabolically with t. Near the leading edge $x = 0$ a further non-uniformity will occur, and for details of the inner region needed to deal with this reference may be made to Van Dyke (1964).

The existence of the similarity solution, and the choice of similarity variable, may at first sight appear surprising but it is the absence of any length scale L which gives the special structure. For the flow of a stream past a smooth two-dimensional curved surface, curvilinear coordinates x, y may be taken so that $y = 0$ is the surface and $x = 0$ the leading edge. The boundary-layer equations will be eqns (5.19), since all the additional terms due to the surface curvature will be $O(1/Re)$. However, the outer solution will no longer be a uniform flow, the pressure will vary with x, and it must be retained in the boundary-layer problem. In general this will introduce a length scale in the x direction, and similarity solutions do not exist. An exception is for the problem of flow past a wedge, and this is discussed in Exercise 5.3b.

Boundary layers also commonly occur in convective heat flow, and we have discussed in Chapter 4 the steady flow of an inviscid stream past a semi-infinite heated flat plate. If, however, the flow is viscous it will in general have a velocity which is not parallel to the plate, so that the energy equation for the temperature T, assuming that the fluid is incompressible, is

$$\frac{\partial T}{\partial t} + u\frac{\partial T}{\partial x} + v\frac{\partial T}{\partial y} = \frac{1}{Pe}\left(\frac{\partial^2 T}{\partial x^2} + \frac{\partial^2 T}{\partial y^2}\right), \qquad (5.21)$$

where Pe is the Peclet number and the velocity components u and v are determined from the momentum and continuity equations for the fluid. For large Peclet numbers we expect that a boundary layer will occur, because with $1/Pe = 0$ the terms containing the highest derivatives in eqn (5.21) vanish.

If the problem is steady and there is a Poiseuille flow between two fixed walls $y = 0$ and $y = 1$, $-\infty < x < \infty$, then the stream function ψ satisfies $d^4\psi/dy^4 = 0$, with $\psi = d\psi/dy = 0$ on $y = 0$; $\psi = 1$, $\partial\psi/\partial y = 0$ on $y = 1$; the imposed pressure gradient has been chosen so that the mass flow down the channel is unity. Then $\psi = 3y^2 - 2y^3$, and if the section of the wall $y = 0$, $x > 0$ has unit temperature with the remainder of the walls at zero temperature, then T satisfies the boundary-value problem

$$6(y - y^2)\frac{\partial T}{\partial x} = \frac{1}{Pe}\left(\frac{\partial^2 T}{\partial x^2} + \frac{\partial^2 T}{\partial y^2}\right), \qquad (5.22)$$

where $T = 1$, $y = 0$, $x > 0$; $T = 0$, $y = 0$, $x < 0$; $T = 0$, $y = 1$, $-\infty < x < \infty$; and $T \to 0$ as $x \to -\infty$, $0 < y < 1$. The outer solution for $Pe \gg 1$ is clearly $T = 0$, which does not satisfy the condition on

$y = 0$, $x > 0$. We therefore look for a boundary layer adjacent to $y = 0$ in which x and T are both $O(1)$. The scaling $\bar{y} = (Pe)^{\frac{1}{3}}y$ gives an appropriate balance of terms, and

$$6\bar{y}\frac{\partial T}{\partial x} = \frac{\partial^2 T}{\partial \bar{y}^2} + O(Pe^{-\frac{1}{3}}). \tag{5.23}$$

The boundary condition is $T = 1$, $\bar{y} = 0$, $x > 0$; the matching conditions are $T \to 0$ as $\bar{y} \to \infty$ together with some statement that the oncoming flow has zero temperature, such as $T = 0$, $x = 0$, $\bar{y} > 0$. Note that the non-uniform velocity u has given rise to a thicker boundary layer $O(Pe^{-\frac{1}{3}})$ than in the constant-velocity case where it is $O(Pe^{-\frac{1}{2}})$, with consequently larger error terms. Also we may expect that there will be some leading-edge non-uniformity, and the condition $T = 0$ at $x = 0$ will not be satisfied for small enough $\bar{y}$.

Ignoring terms $O(Pe^{-\frac{1}{3}})$ the problem has a similarity solution in the variable $\eta = \bar{y}/x^{\frac{1}{3}}$ where T satisfies

$$\frac{d^2 T}{d\eta^2} + 2\eta^2 \frac{dT}{d\eta} = 0,$$

$$T(0) = 1, \quad T(\infty) = 0.$$

This may be integrated to give

$$T = A\int_{y(Pe/x)^{\frac{1}{3}}}^{\infty} e^{-2s^3/3}\, ds, \tag{5.24}$$

where

$$A = \int_{0}^{\infty} e^{-2s^3/3}\, ds.$$

The boundary layer exists when $y \sim (x/Pe)^{\frac{1}{3}}$ as in Fig. 5.8. Non-uniformities will occur far downstream when $x \sim Pe$ and the thickness is comparable with the channel width, and also at the leading edge. The leading-edge non-uniformity will occur when x is small enough so that all three terms in eqn (5.22) are of the same order of magnitude, and it is easily verified that this requires that x and y are $O(Pe^{-\frac{1}{2}})$.

A temperature boundary layer may also occur in the flow of a slightly viscous fluid past a semi-infinite heated flat plate, depending on the relative sizes of Pe and Re. In the scaled variables used

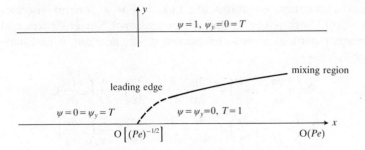

Fig. 5.8. Temperature boundary layer in a channel

in the fluid boundary-layer equations (5.19) and (5.20), eqn (5.21) for steady flow becomes

$$\frac{Pr}{2} f(\eta) \frac{\mathrm{d}T}{\mathrm{d}\eta} + \frac{\mathrm{d}^2 T}{\mathrm{d}\eta^2} = 0,$$

where Pr is the Prandtl number Pe/Re. Hence if the Prandtl number is $O(1)$ and $Re \sim Pe$, then

$$T = \int_{\eta}^{\infty} \exp\left[-\frac{Pr}{2} \int^{s} f(s')\, \mathrm{d}s' \right] \mathrm{d}s,$$

where f is defined by eqn (5.20) and called the Blasius function. For large values of Pr the boundary layer is where η is small and $f \sim c\eta^2$, since $f(0) = f'(0) = 0$. In this situation η must be rescaled with $\bar{\eta} = (Pr)^{\frac{1}{3}}\eta$ to give

$$\frac{c\bar{\eta}^2}{2} \frac{\mathrm{d}T}{\mathrm{d}\bar{\eta}} + \frac{\mathrm{d}^2 T}{\mathrm{d}\bar{\eta}^2} = 0,$$

describing a temperature boundary layer similar to that given by eqn (5.24). In physical terms with $Pr \gg 1$ and $Re \ll 1$ there is a shear flow in the temperature boundary layer which is much thinner than the fluid boundary layer, although both grow as $x^{\frac{1}{2}}$.

A more complicated situation occurs when the fluid viscosity is temperature-dependent, since the fluid velocity will now vary with the temperature profile in the boundary layer and the energy and momentum equations are linked. For a general viscosity $\mu = \mu(T)$ with $\mu(0) = 1$, the non-dimensional Navier–Stokes and energy equations in two-dimensional steady flow are, from Chapter 1,

$$Re\left(u\frac{\partial u}{\partial x} + v\frac{\partial u}{\partial y}\right) = -\frac{\partial p}{\partial x} + \frac{\partial}{\partial x}\left(\mu\frac{\partial u}{\partial x}\right) + \frac{\partial}{\partial y}\left(\mu\frac{\partial u}{\partial y}\right),$$

$$Re\left(u\frac{\partial v}{\partial x} + v\frac{\partial v}{\partial y}\right) = -\frac{\partial p}{\partial y} + \frac{\partial}{\partial x}\left(\mu\frac{\partial v}{\partial x}\right) + \frac{\partial}{\partial y}\left(\mu\frac{\partial v}{\partial y}\right),$$

$$Pe\left(u\frac{\partial T}{\partial x} + v\frac{\partial T}{\partial y}\right) = \frac{\partial^2 T}{\partial x^2} + \frac{\partial^2 T}{\partial y^2} + O(Br),$$

$$\frac{\partial u}{\partial x} + \frac{\partial v}{\partial y} = 0.$$

Here the pressure has been made non-dimensional with the viscous pressure $U\mu_0/L$, and the dissipation terms in the energy equation are neglected ($Br \ll 1$).

For a flow in a channel $0 < y < 1$ with walls at uniform temperature $T = 0$, the Poiseuille-flow solution of these equations is

$$T = 0, \quad \psi = 3y^2 - 2y^3,$$

where ψ is the stream function made non-dimensional such that $\psi = 1$ on $y = 1$, that is unit mass-flow down the channel. If the section of the wall $y = 0$, $x > 0$ is heated to temperature $T = 1$ then a temperature boundary layer will form if the Peclet number is large. The scaling is not obvious and we write $\bar{y} = y/\varepsilon$, where ε is to be chosen in terms of Pe. The temperature and x scale remain $O(1)$, and the matching gives that as $\bar{y} \to \infty$, $T \to 0$ and $\psi \sim 3y^2$. The appropriate scaling for $\bar{\psi}$ is therefore $\bar{\psi} = \psi/\varepsilon^2$ so that as $\bar{y} \to \infty$, $\bar{\psi} \sim 3\bar{y}^2$. Substituting this scaling in the equations, and assuming slow flow so that $Re = 0$,

$$\frac{\partial}{\partial \bar{y}}\left(\mu\frac{\partial^2 \bar{\psi}}{\partial \bar{y}^2}\right) = \varepsilon\frac{\partial p}{\partial x}, \quad \frac{\partial p}{\partial y} = O(\varepsilon),$$

$$\varepsilon^2\frac{\partial^2 T}{\partial x^2} + \frac{\partial^2 T}{\partial \bar{y}^2} = \varepsilon^3 Pe\left(\frac{\partial \bar{\psi}}{\partial \bar{y}}\frac{\partial T}{\partial x} - \frac{\partial \psi}{\partial x}\frac{\partial T}{\partial \bar{y}}\right).$$

Clearly we choose $\bar{p} = \varepsilon p$ and $\varepsilon = (Pe)^{-\frac{1}{3}}$ as in the case $\mu = 1$ to obtain

$$\frac{\partial^2}{\partial \bar{y}^2}\left[\mu(T)\frac{\partial^2 \psi}{\partial \bar{y}^2}\right]=0,$$

$$\frac{\partial^2 T}{\partial \bar{y}^2}=\frac{\partial \bar{\psi}}{\partial \bar{y}}\frac{\partial T}{\partial x}-\frac{\partial \bar{\psi}}{\partial x}\frac{\partial T}{\partial \bar{y}}. \tag{5.25}$$

A similarity solution may be found with $\eta = \bar{y}/x^{\frac{1}{3}}$, $T = T(\eta)$, $\bar{\psi} = x^{\frac{2}{3}}f(\eta)$, where the form of T and $\bar{\psi}$ is dictated by the boundary conditions $T = 1$ on $\eta = 0$ and $\bar{\psi} \sim 3x^{\frac{2}{3}}\eta^2$ as $\eta \to \infty$. Then eqns (5.25) become, with two integrations,

$$\mu(T)\frac{d^2 f}{d\eta^2}=6,$$

$$\frac{d^2 T}{d\eta^2}+\tfrac{2}{3}f\frac{dT}{d\eta}=0. \tag{5.26}$$

The boundary conditions on this pair of second-order non-linear equations are

$$\eta = 0, \quad f = 0 = \frac{df}{d\eta}, \quad T = 1; \quad \eta \to \infty, \quad T \to 0.$$

The boundary-value problem has similarity properties and may readily be solved numerically. If $\mu = 1$ the solution reduces to eqn (5.24). This more complicated boundary layer still grows as $x^{\frac{1}{3}}$ and is valid for $x \ll Pe$ and away from the leading edge, having qualitatively similar properties to that shown in Fig. 5.8.

An alternative view of this problem is to consider a length scale L which is $Pe \times (\text{width})$ so that the channel is narrow, and to heat both walls for $x > 0$; there will then be identical boundary layers growing on each wall for $x \ll 1$, and they will mix when x is $O(1)$.

The effects of dissipation can also give rise to interesting boundary layers, and we consider first the case of constant viscosity $\mu = 1$ with large Brinkman number such that $Br \sim Pe$, and by scaling the temperature we may take $Br = Pe$. The flow in the channel is given by $\psi = 3y^2 - 2y^3$ as before, so that the energy equation away from any steep regions reduces to

$$y(1-y)\frac{\partial T}{\partial x}=6(1-2y)^2.$$

This integrates to give

$$T = \frac{6x(1-2y)^2}{y(1-y)}, \tag{5.27}$$

which is singular at the walls so that no boundary condition on T can be satisfied. If for simplicity we require that $T = 0$ at the walls, then a boundary layer will occur and the conduction term $(1/Pe)(\partial^2 T/\partial y^2)$ must balance with the convection and dissipation terms. For a balance between conduction and convection a boundary layer of thickness $O(Pe^{-\frac{1}{3}})$ is required as before, with $\bar{\psi} = (Pe)^{\frac{2}{3}}\psi$ and $\bar{\psi} \sim 3\bar{y}^2$ as $\bar{y} \to \infty$. In addition T must be scaled and must match with $6x/y = (6x/\bar{y})(Pe)^{\frac{1}{3}}$ as $\bar{y} \to \infty$ so that $\bar{T} = T(Pe)^{-\frac{1}{3}}$ in this layer. The energy equation now becomes

$$\frac{\partial \bar{\psi}}{\partial \bar{y}}\frac{\partial T}{\partial x} - \frac{\partial \bar{\psi}}{\partial x}\frac{\partial \bar{T}}{\partial \bar{y}} = \frac{\partial^2 \bar{T}}{\partial \bar{y}^2} + \left(\frac{\partial^2 \bar{\psi}}{\partial \bar{y}^2}\right)^2, \tag{5.28}$$

where $\partial^2 \bar{\psi}/\partial \bar{y}^2 = 1$ from the momentum equation.

This has a similarity solution with $\bar{\psi} = x^{\frac{2}{3}}f(\eta)$, $\bar{T} = x^{\frac{2}{3}}g(\eta)$, $\eta = \bar{y}/x^{\frac{1}{3}}$, where the form of $\bar{T}$ is also chosen by the matching condition $\bar{T} \sim 6x^{\frac{2}{3}}/\eta$ as $\eta \to \infty$. Equation (5.28) reduces to

$$\frac{d^2 g}{d\eta^2} + \frac{\eta^2}{3}\frac{dg}{d\eta} - \frac{2\eta}{3}g = -1, \tag{5.29}$$

with boundary conditions $\eta = 0$, $g = 0$; $\eta \to \infty$, $g \sim 6/\eta$. In this boundary layer T has a large maximum $O(Pe^{\frac{1}{3}})$, its thickness grows as $x^{\frac{1}{3}}$, and the result will be valid for $x \ll Pe$. A typical temperature profile is shown in Fig. 5.9. Physically almost all the heat dissipated is contained in two spikes, and the Poiseuille flow is unaffected.

In Chapter 3 we noted that when dissipation was important and the viscosity varied exponentially with temperature so that $\mu = e^{-\beta\tau}$, where β is a non-dimensional parameter called the Nahme–Griffith number, then a steady flow in an infinite channel did not exist if β was large enough. In the unsteady case this led to the idea of thermal runaway, and this phenomenon is likely to occur in the steady flow down a semi-infinite channel if the pressure gradient or mass flow is large enough. If we examine the above boundary-layer problem for flow with dissipation and

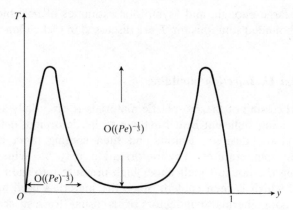

Fig. 5.9. Temperature profile in a channel with dissipation

$\mu = e^{-\beta T}$ where $\beta \ll 1$, then outside the boundary layer where T is $O(1)$, $\mu = 1 + O(\beta)$. In the layer, T is large and $\mu = \exp[-\beta(Pe)^{\frac{1}{3}}\bar{T}]$ so that for $\beta \sim (Pe)^{-\frac{1}{3}}$ the viscosity is not constant in the layer. With the same scaling the momentum and energy equations become, with $\beta = b(Pe)^{-\frac{1}{3}}$,

$$\frac{\partial^2 \bar{\psi}}{\partial \bar{y}^2} = e^{b\bar{T}},$$

$$\frac{\partial \bar{\psi}}{\partial \bar{y}}\frac{\partial \bar{T}}{\partial x} - \frac{\partial \bar{\psi}}{\partial x}\frac{\partial \bar{T}}{\partial \bar{y}} = \frac{\partial^2 \bar{T}}{\partial \bar{y}^2} + e^{b\bar{T}}. \qquad (5.30)$$

The boundary conditions are, as before,

$$\bar{y} = 0, \quad \bar{\psi} = 0 = \frac{\partial \bar{\psi}}{\partial \bar{y}}, \quad \bar{T} = 0; \quad \bar{y} \to \infty, \quad \bar{\psi} \sim \tfrac{1}{2}\bar{y}^2, \quad \bar{T} \sim \frac{\bar{x}}{\bar{y}};$$

together with matching conditions as $\bar{x} \to 0$. There is now no similarity solution for this boundary-value problem, and a numerical solution is not at all easy to obtain. The energy equation does, however, bear a strong resemblance to the equation of unsteady flows in an infinite length channel, since its left hand side is derived from the total time derivative dT/dt. Hence we may conjecture that it will exhibit blow-up after a finite distance

Asymptotic methods for non-linear problems

if b is large enough, and asymptotic estimates of possible forms for unbounded solutions for T are discussed in Ockendon (1979).

Problem U. *Injection moulding*

The processing of thermo-plastic materials is essentially a matter of providing sufficient heat for them to be deformed, deforming them in the desired fashion, and then cooling them back to ambient temperatures. The deformation may be achieved by injecting the hot soft melt under high pressure into the required mould, which is then cooled. Since the materials used are normally poor thermal conductors it is usual to restrict mould geometries to thin layers, so that the cooling time is short enough for the mould to be used repeatedly. The process can easily be automated if the injection and cooling times can be estimated. The quality of the product is important, and in particular the physical properties of the moulded object near its surface need careful control. These physical properties will in general depend on the deformation and temperature history of the local material element, and a mathematical model for the process is very desirable if a high rejection rate is not to occur. Thus, for example, in the moulding of a plastic lens whose surface is to be polished, high stress-gradients must be avoided near the surfaces as the mould cools, or the polishing procedure will produce cracks in the lens.

The physical properties of thermoplastics in their molten and semi-molten states are extremely complicated, and attempts to construct appropriate constitutive relations between stress and strain are necessarily semi-empirical. However, some understanding of the flow and heat transfer in the injection-moulding process may be obtained from a model which assumes incompressible viscous fluid flow, with a viscosity which depends exponentially on temperature. To further simplify the problem we only consider moulds with two parallel plane walls which are a small distance apart. A typical situation is shown in Fig. U1, where the molten plastic is injected at the origin between planes $z = 0$ and $z = 1$ in non-dimensional variables. The air inside the mould escapes through blowholes, and the positioning of the blowholes and injection point is an important design consideration.

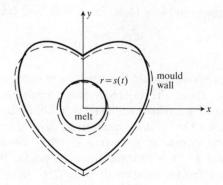

Fig. U1. Injection moulding

A simple model for the flow is that of slow flow in a narrow uniform channel (Hele–Shaw flow), but with viscosity $\mu = e^{-\beta(T-1)}$. Modifying the analysis leading to eqns (3.9) we have

$$\frac{\partial p}{\partial z} = 0 = \frac{\partial u}{\partial x} + \frac{\partial v}{\partial y} + \frac{\partial w}{\partial z}, \tag{U1}$$

also

$$\frac{\partial p}{\partial x} = \frac{\partial}{\partial z}\left(e^{-\beta(T-1)}\frac{\partial u}{\partial z}\right),$$

$$\frac{\partial p}{\partial y} = \frac{\partial}{\partial z}\left(e^{-\beta(T-1)}\frac{\partial v}{\partial z}\right), \tag{U2}$$

together with the energy equation

$$Pe^*\left(\frac{\partial T}{\partial t} + u\frac{\partial T}{\partial x} + v\frac{\partial T}{\partial y} + w\frac{\partial T}{\partial z}\right) = \frac{\partial^2 T}{\partial z^2}, \tag{U3}$$

where $Pe^* = \rho(ULc/k)(h/L)^2$. Although the flow is slow there will be temperature variations in time due to the effects of the moving fluid front, and the derivative $\partial T/\partial t$ can not be neglected. Dissipation effects have been neglected for simplicity, but will be re-introduced later. If we now assume that the modified Peclet number Pe^* is large, then the fluid temperature is constant, equal to that of the hot inlet fluid $T = 1$, except for boundary layers at the walls and at the moving front. Thus in the core the equations reduce to eqns (3.9) for a Hele–Shaw flow, and from eqn (3.10)

$$\frac{\partial^2 p}{\partial x^2} + \frac{\partial^2 p}{\partial y^2} = 0.$$

In a two-dimensional mould there will be two stages; a symmetrical expansion about the injection point until the moving front meets the mould boundary, followed by a filling stage until all the air is expelled. The initial symmetrical flow is easily obtained since from the symmetry $p = -f(t)\log[r/s(t)]$, where the moving front is $r = s(t)$ and r is distance from the origin. If the injection mass-flow rate (non-dimensional) is $2\pi\dot{Q}(t)$, then $f(t) = 12\dot{Q}(t)$. Finally the kinematic condition of eqn (3.11) gives that $f = 12s\dot{s}$ and $s^2 = 2Q(t)$, a result which could have been obtained directly by an overall mass balance. The subsequent filling stage is much more difficult, since there will be moving boundaries on which $p = 0$ and eqn (3.11) is satisfied, and fixed boundaries on which $\partial p/\partial n = 0$. This is a difficult numerical problem, and we shall only consider one-dimensional moulds in which there is no filling stage. In the one-dimensional case with mass flow $\dot{Q}(t)$,

$$p = -12\dot{Q}(t)(x - s)$$

where $s = Q(t)$, so that for constant mass-flow rate we have $s \sim t$, whereas for constant inlet pressure $s \sim t^{\frac{1}{2}}$. The non-dimensional injection and filling time t_0 is given by $Q(t_0) = 1$ if the length of the mould was L.

There is also a third packing stage in which the molten material is compressed, so that on cooling there is a pressure reduction rather than a contraction. For a model for this stage see Exercise 5.3c.

Since surface conditions are so important we focus attention on the boundary layers which form there, and we consider the one-dimensional mould. On the time scale for filling the mould, t is $O(1)$ so that the boundary-layer thickness is $O[(Pe^*)^{-\frac{1}{2}}]$ and its 'leading edge' is at the moving front $x = s(t)$, as shown in Fig. U2. With the scaling $\bar{z} = z(Pe^*)^{\frac{1}{2}}$, and recalling that $u \sim z$ in the layer, eqn (U3) reduces to

$$\frac{\partial T}{\partial t} = \frac{\partial^2 T}{\partial \bar{z}^2}, \tag{U4}$$

subject to boundary conditions $T = 0$, $\bar{z} = 0$, $x < s(t)$; $T = 0$,

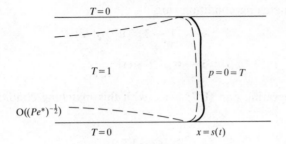

Fig. U2. One-dimensional mould

$x = s(t), t > 0$, and matching condition $T = 1, \bar{z} \to \infty$. In axes moving with the fluid front $x = s(t)$, which may be rewritten $t = \bar{s}(x)$, this is just a Rayleigh problem, with a similarity solution given by eqn (5.16), which in our variables gives

$$T = \operatorname{erf}\left(\frac{\bar{z}}{2[t - \bar{s}(x)]^{\frac{1}{2}}}\right).$$ (U5)

This represents a boundary layer trailing from the moving front which grows as the square root of the time elapsed since the moving front passed any given point x. It will not be valid at the leading edge $x = s(t)$, nor near the inlet $x = 0$ where $T = 1$ for all $z > 0$, so that the boundary thickness must be zero there. We examine this non-uniformity near the inlet by rescaling x and u so that in eqn (U3) the term $u(\partial T/\partial x)$ may not be neglected in comparison with $\partial T/\partial t$. Also we require that $\bar{z}$ and T are O(1) so that we can match T with $\operatorname{erf}(\bar{z}/2\sqrt{t})$ as $\bar{x} \to \infty$, and finally we must choose $\bar{u}$ so that it matches with the mainstream or core as $\bar{z} \to \infty$. In this unsteady problem the core flow depends on time and $\psi = \dot{Q}(t)(3z^2 - 2z^3)$; hence the matching conditions as $\bar{z} \to \infty$ are $u \sim 6\dot{Q}z$, $w \sim 0$. The only consistent scaling is $\bar{x} = x(Pe^*)^{\frac{1}{2}}$, $\bar{u} = u(Pe^*)^{\frac{1}{2}}$, $\bar{w} = w(Pe^*)^{\frac{1}{2}}$, and eqns (U1), (U2) and (U3) become

$$\frac{\partial \bar{u}}{\partial \bar{x}} + \frac{\partial \bar{w}}{\partial \bar{z}} = 0,$$ (U6)

$$\frac{\partial^2}{\partial \bar{z}^2}\left(e^{-\beta(T-1)}\frac{\partial \bar{u}}{\partial \bar{z}}\right) = 0,$$ (U7)

$$\frac{\partial T}{\partial t} + \bar{u}\frac{\partial T}{\partial \bar{x}} + \bar{w}\frac{\partial T}{\partial \bar{z}} = \frac{\partial^2 T}{\partial \bar{z}^2}.$$ (U8)

The matching conditions are

$$\bar{z} \to \infty, \; T \to 1, \quad \bar{u} \sim 6\dot{Q}\bar{z};$$

$$\bar{x} \to \infty, \quad T \sim \text{erf} \frac{\bar{z}}{2\sqrt{t}};$$

and integrating eqn (U7) twice with this matching condition we obtain

$$\frac{\partial \bar{u}}{\partial \bar{z}} = 6\dot{Q}e^{\beta(T-1)}. \tag{U9}$$

The boundary conditions are $T = 0 = \bar{u} = \bar{w}$, $\bar{z} = 0$, $t > 0$, $\bar{x} > 0$, together with further conditions either at $t = 0$, $x > 0$ or $x = 0$, $t > 0$. This is now a problem in three independent variables, and it is not a simple matter to determine in general what boundary conditions are needed for a well-posed problem. However, a similarity transformation exists which reduces the equations and their necessary boundary conditions to a problem in two independent variables if $\dot{Q}$ is proportional to a power of t. For simplicity we discuss the case of constant inlet pressure, so that (say) $\dot{Q} = c/6t^{\frac{1}{2}}$, and constant viscosity $\beta = 0$. Then defining $\zeta = \bar{z}/2\sqrt{t}$ and $\xi = x/2ct$, eqn (U9) gives $\bar{u} = 2c\zeta$ and, from eqn (U6), $\bar{w} = 0$. Finally eqn (U8) reduces to

$$\frac{\partial^2 T}{\partial \zeta^2} + 2\zeta \frac{\partial T}{\partial \zeta} = 4(\zeta - \xi) \frac{\partial T}{\partial \xi}, \tag{U10}$$

with boundary conditions

$$T = 1, \quad \zeta \to \infty, \quad \xi > 0;$$

$$T = 0, \quad \zeta = 0, \quad \xi > 0; \tag{U11}$$

$$T \to \text{erf} \, \zeta, \quad \xi \to \infty, \quad \zeta > 0.$$

These conditions are shown schematically in Fig. U3a and are appropriate for a 'backward diffusion' equation, that is for $\zeta < \xi$. However, in the domain $\zeta > \xi$ we would expect that initial data would be needed to define the solution, and this corresponds to presenting data in the physical problem on $\bar{x} = 0$ or as $t \to \infty$. In either case a sensible physical condition would seem to be $T = 1$ except possibly for small enough $\bar{z}$, so that on $\xi = 0$, $\zeta > 0$, $T = 1$.

This is an unusual boundary-value problem for a parabolic

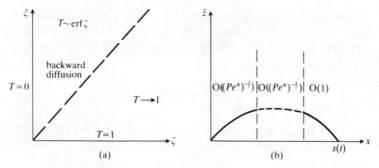

Fig. U3. Domains for temperature boundary value problem

equation because the diffusion coefficient changes sign (and is unbounded at $\zeta = \xi$), and no firm results are available about its well-posedness. However, numerical schemes can be devised in terms of $\theta = T - \text{erf } \eta$, so that $\theta \to 0$ as $\zeta \to \infty$ and $\xi \to \infty$, which appear to give a unique solution to the problem (see for example Tayler and Nicholas (1982)). There is a singularity at $\xi = \zeta = 0$ which may be resolved by changing to variables ξ and $\eta = \zeta(2/3\xi)^{\frac{1}{3}}$, so that for small ξ

$$\frac{\partial^2 T}{\partial \eta^2} + 2\eta^2 \frac{\partial T}{\partial \eta} \sim 0$$

with $T(0) = 0$, $T(\infty) = 1$. This is identical to the problem of the steady flow of a stream past a cooled semi-infinite flat plate solved by eqn (5.24), and the numerical solution of eqn (U10) subject to eqns (U11) gives a transition from this well-known thermal entry layer to a layer trailing behind the moving front as in Fig. U3b. The properties of this 'confined' boundary layer may be examined for varying values of β and different entry mass-flow rates, and maximum values for the temperature and pressure gradients estimated.

If the Brinkman number is not small then the above confined boundary layer will not be valid for all t, and dissipation effects may become important before the mould is filled.

5.4. Far-field solutions

The WKB method may be generalized to linear partial differential equations, and may be used either to obtain an irregular

expansion in a small parameter or as an asymptotic expansion for
large values of the independent variables. Thus for the wave
equation in two space variables, solutions of frequency ω satisfy
the Helmholz equation

$$\frac{\partial^2 u}{\partial x^2} + \frac{\partial^2 u}{\partial y^2} + \omega^2 u = 0. \tag{5.31}$$

For high frequencies, and in non-dimensional variables with ω
replaced by $1/\varepsilon$, $\varepsilon \ll 1$, a WKB expansion is sought in the form

$$u = \exp i\left[\frac{1}{\varepsilon} g_0(x, y) + g_1 + O(\varepsilon)\right], \tag{5.32}$$

where i is introduced for convenience.
  Substituting in eqn (5.31) and equating powers of ε^{-2} and ε^{-1},

$$\left(\frac{\partial g_0}{\partial x}\right)^2 + \left(\frac{\partial g_0}{\partial y}\right)^2 = 1, \tag{5.33}$$

$$\frac{\partial g_0}{\partial x}\frac{\partial g_1}{\partial x} + \frac{\partial g_0}{\partial y}\frac{\partial g_1}{\partial y} + \frac{1}{2}\left(\frac{\partial^2 g_0}{\partial x^2} + \frac{\partial^2 g_0}{\partial y^2}\right) = 0, \tag{5.34}$$

where eqn (5.33) is called the *eikonal equation* and is widely
used in geometrical optics. The second-order linear equation
(5.32) has been asymptotically replaced by a sequence of first-
order equations (5.33), (5.34), etc., the first of which is non-
linear. There is a considerable body of knowledge about non-
linear first-order equations, and eqn (5.33) is simpler to solve
than the full problem of eqn (5.31). Further terms in the asymptotic expansion may, however, be extremely tedious to calculate
and usually this WKB procedure is only used to obtain at most
two terms.
  A minor variant of the method is to look for an expansion in
the form

$$u = A(x, y; \varepsilon)\exp\left[\frac{i}{\varepsilon} g(x, y)\right],$$

where A is assumed to be regular in ε. Then g satisfies eqn (5.33)
and if $A = A_0(x, y) + O(\varepsilon)$, A_0 satisfies

$$\frac{\partial g}{\partial x}\frac{\partial A_0}{\partial x} + \frac{\partial g}{\partial y}\frac{\partial A_0}{\partial y} + \frac{A_0}{2}\left(\frac{\partial^2 g}{\partial x^2} + \frac{\partial^2 g}{\partial y^2}\right) = 0. \tag{5.35}$$

It is common to use the terminology of amplitude for A and phase for g, although in this context they are functions of the independent variables. The method may also be used to find the asymptotic form of the solution for large values of the independent variables; it is conveniently reduced to an expansion in a smaller parameter ε by rescaling the large independent variables with ε, as in the use of the WKB method for ordinary differential equations described in Section 5.1a of this chapter. Such asymptotic results are called *far-field solutions*, and the method is not restricted to two independent variables.

Thus for small-amplitude gravity waves on the surface of a large expanse of water of depth h, the boundary-value problem is given by eqn (1.30), where in non-dimensional variables we may take $g = 1$, and h is constant or slowly varying. A far-field solution is found by writing $\varepsilon x = \bar{x}$, $\varepsilon y = \bar{y}$, and assuming that $h = h(\bar{x}, \bar{y})$. Since any disturbance travels with speed which is O(1), we must also rescale time by $\varepsilon t = \bar{t}$ so that the far-field solution also implies large time. Then

$$\frac{\partial^2 \phi}{\partial \bar{x}^2} + \frac{\partial^2 \phi}{\partial \bar{y}^2} + \frac{1}{\varepsilon^2}\frac{\partial^2 \phi}{\partial z^2} = 0;$$

$$z = -h, \quad \frac{\partial \phi}{\partial z} = 0; \quad z = 0, \quad \frac{\partial^2 \phi}{\partial t^2} + \frac{1}{\varepsilon^2}\frac{\partial \phi}{\partial z} = 0.$$

An expansion $\phi = A(x, y, z; \varepsilon)\exp[(i/\varepsilon)g(x, y, t)]$ is possible provided g is independent of z, and

$$-\left[\left(\frac{\partial g}{\partial \bar{x}}\right)^2 + \left(\frac{\partial g}{\partial \bar{y}}\right)^2\right]A + \frac{\partial^2 A}{\partial z^2} = 0;$$

together with

$$z = -h, \quad \frac{\partial A}{\partial z} = 0; \quad z = 0, \quad \frac{\partial A}{\partial z} = \left(\frac{\partial g}{\partial \bar{t}}\right)^2 A.$$

This is an ordinary differential equation for A in the variable z, and is an eigenvalue problem. A non-trivial solution only exists if

$$\left[\left(\frac{\partial g}{\partial \bar{x}}\right)^2 + \left(\frac{\partial g}{\partial \bar{y}}\right)^2\right]^{\frac{1}{2}}\tanh\left\{h\left[\left(\frac{\partial g}{\partial \bar{x}}\right)^2 + \left(\frac{\partial g}{\partial \bar{y}}\right)^2\right]^{\frac{1}{2}}\right\} = \left(\frac{\partial g}{\partial \bar{t}}\right)^2,$$

$$(5.36)$$

and then

$$A = \alpha \cosh\left\{(z+h)\left[\left(\frac{\partial g}{\partial \bar{x}}\right)^2 + \left(\frac{\partial g}{\partial \bar{y}}\right)^2\right]^{\frac{1}{2}}\right\}.$$

For large h, eqn (5.36) reduces to the non-linear first-order equation

$$\frac{\partial g}{\partial \bar{t}} = \left[\left(\frac{\partial g}{\partial \bar{x}}\right)^2 + \left(\frac{\partial g}{\partial \bar{y}}\right)^2\right]^{\frac{1}{4}}, \qquad (5.37)$$

where either real root may be appropriate.

In Chapter 2, Section 2.6 simple solutions of linear hyperbolic equations with constant coefficients and associated boundary conditions were obtained by looking for solutions which were periodic both in time and in the unbounded space variables. Thus for water waves a solution is

$$\phi = f(z)e^{i(lx + my - \omega t)},$$

where $f(z) = \alpha \cosh[(z+h)(l^2 + m^2)^{\frac{1}{2}}]$, provided that the wave numbers satisfy a dispersion relation $\omega = F(l, m)$. In the case of waves on water of infinite depth the dispersion relation is $\omega = \pm(l^2 + m^2)^{\frac{1}{4}}$.

In obtaining a dispersion relation, the operation of differentiation with respect to x in the boundary-value problem has been replaced by multiplication by il, with respect to y by im, and with respect to t by $-i\omega$. In the WKB procedure for obtaining the equation satisfied by the phase function g, derivatives higher than the first in the full problem have been replaced by squares and higher powers of the first derivative. Thus the identical operations are performed either in evaluating a dispersion relation, or in obtaining the equation satisfied by g. If we therefore associate l with $\partial g/\partial x$, m with $\partial g/\partial y$ and $-\omega$ with $\partial g/\partial t$, the dispersion relation $\omega = F(l, m)$ for a given problem may be used to give the equation for the phase function of the far-field solution in the form

$$\frac{\partial g}{\partial t} + F\left(\frac{\partial g}{\partial x}, \frac{\partial g}{\partial y}\right) = 0. \qquad (5.38)$$

In the case of water waves this may immediately be verified by eqn (5.37); note that F may depend explicitly on x, y or t, if for example h is slowly varying in time and space.

Before discussing the general solution of non-linear first-order equations we consider the simple case of only two independent variables, say t and x, and F independent of x and t. Then if $\omega(x, t) = -\partial g/\partial t$ and $k(x, t) = \partial g/\partial x$, the dispersion relation is $\omega = F(k)$ and

$$\frac{\partial k}{\partial t} = \frac{\partial^2 g}{\partial x\, \partial t} = -\frac{\partial \omega}{\partial x} = -F'(k)\frac{\partial k}{\partial x}. \tag{5.39}$$

This is a very simple first-order partial differential equation for k with general solution

$$k = H[x - F'(k)t],$$

where H is an arbitrary function. On lines of slope $F'(k)$ in the x, t plane (or, recognizing that t is time, lines of speed $F'(k)$), k is constant and $F'(k)$ is called the group velocity of the waves. Thus in the far field the phase function $g = kx - F(k)t$, where k is defined by $x = F'(k)t$, and the asymptotic form for u is implicitly defined. The lines of constant k are the characteristics of the first-order partial differential equation (5.39) and their 'slope' is the group velocity $F'(k)$. For a steady-state problem in two space variables the dispersion relation will be $F(l, m) = 0$, and the equation for $l = \partial g/\partial x$ is

$$\frac{\partial F}{\partial m}\frac{\partial l}{\partial y} + \frac{\partial F}{\partial l}\frac{\partial l}{\partial x} = 0,$$

so that lines of constant l and m will have slope

$$\frac{\partial y}{\partial x} = \left(\frac{\partial F}{\partial m}\right) \Big/ \left(\frac{\partial F}{\partial l}\right), \tag{5.40}$$

and this slope may be interpreted as a 'group velocity'.

An illustrative example is the ship-wave problem in which a point disturbance moves steadily with unit speed (in the negative x direction say) on the surface of water of infinite depth. In a frame moving with the ship, time derivatives are then replaced by $(\partial/\partial x)$, and ω in the dispersion relation is replaced by l. It then becomes $l = (l^2 + m^2)^{\frac{1}{4}}$, and g satisfies an equation similar to (5.37), namely

$$\frac{\partial g}{\partial x} - \left[\left(\frac{\partial g}{\partial x}\right)^2 + \left(\frac{\partial g}{\partial y}\right)^2\right]^{\frac{1}{4}} = 0. \tag{5.41}$$

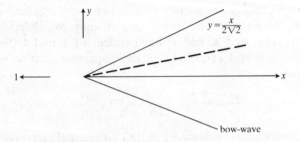

Fig. 5.10. Ray diagram for ship wave problem

From eqn (5.40) l and m will be constant, equal to $k \cos \theta$ and $k \sin \theta$ for convenience, on lines of slope

$$\frac{\mathrm{d}y}{\mathrm{d}x} = \frac{-\frac{1}{2}k^{-\frac{1}{2}}\sin \theta}{1 - \frac{1}{2}k^{-\frac{1}{2}}\cos \theta},$$

where $k^{\frac{1}{2}}\cos \theta = 1 \ (k \neq 0)$. Hence

$$\frac{\mathrm{d}y}{\mathrm{d}x} = \frac{\cos \theta \sin \theta}{\cos^2\theta - 2} \quad \text{and} \quad \left|\frac{\mathrm{d}y}{\mathrm{d}x}\right| \leq \frac{1}{2\sqrt{2}}.$$

Thus disturbances which emanate from the origin of the far-field variables x and y, that is some finite neighbourhood of the ship, can only propagate into a wedge-shaped region behind the ship, of semi-angle $\sin^{-1}(\frac{1}{3})$ as in Fig. 5.10. For any given value of $\mathrm{d}y/\mathrm{d}x$ within the wedge there are two possible wave orientations θ and wave numbers k. This wedge-like region is bounded by a ray on which the WKB expansion is not uniformly valid because $(\mathrm{d}/\mathrm{d}\theta)(\mathrm{d}y/\mathrm{d}x) = 0$. The effect of this singularity may be observed in practice as the dominant bow-wave behind a ship. To evaluate g, and hence obtain the far-field solution, it is necessary to eliminate θ between

$$g = \frac{x \cos \theta + y \sin \theta}{\cos^2\theta}$$

and

$$y(\cos^2\theta - 2) = x \cos \theta \sin \theta.$$

This procedure may also be applied to Problem G which was

discussed earlier using Fourier transforms. The dispersion relation is eqn (G5), and with constant $d = 1$ this may be written

$$\left(\frac{\omega^2}{c_1^2} - k^2\right)^{\frac{1}{2}} \tan\left(\frac{\omega^2}{c_1^2} - k^2\right)^{\frac{1}{2}} = \frac{\mu_2}{\mu_1}\left(k^2 - \frac{\omega^2}{c_2^2}\right). \tag{5.42}$$

The substitution $\omega = -\partial g/\partial t$, $k = \partial g/\partial x$ gives the governing equation for g and we are interested in solutions which propagate from $x = 0$ at $t = 0$. Thus $\partial g/\partial x$ will be constant and equal to k on lines $x = (d\omega/dk)t$; and since from Fig. G2, where $d\omega/dk$ and ω/k are plotted against ω, $d\omega/dk$ has a minimum value $G_m > 0$, solutions will only exist for $x/t > G_m$. Also from Fig. G2, $d\omega/dk \leqslant c_2$ for all k so that the propagated signal lies in the region $c_2t \geqslant x \geqslant G_mt$. It is singular on $x = G_mt$ where a larger signal may be expected, and this is in agreement with the results obtained by Fourier transforms. To evaluate the phase function $g = kx - \omega(k)t$ by elimination of k from $x = (d\omega/dk)t$ would not of course be analytically possible in this problem.

A general first-order non-linear partial differential equation in two independent variables has the form

$$F(x, y, u, p, q) = 0, \tag{5.43}$$

where $p = \partial u/\partial x$ and $q = \partial u/\partial y$. Its solution may be obtained from the integral curves of Charpit's equations for the characteristic strips; these may be written, as in Chester (1971),

$$\frac{\partial x}{\partial \tau} = \frac{\partial F}{\partial p}, \quad \frac{\partial y}{\partial \tau} = \frac{\partial F}{\partial q},$$

$$\frac{\partial p}{\partial \tau} = -\frac{\partial F}{\partial x} - p\frac{\partial F}{\partial u},$$

$$\frac{\partial q}{\partial \tau} = -\frac{\partial F}{\partial y} - q\frac{\partial F}{\partial u}, \tag{5.44}$$

$$\frac{\partial u}{\partial \tau} = p\frac{\partial F}{\partial p} + q\frac{\partial F}{\partial q}.$$

Here τ is a variable measured along a given characteristic, and there are five ordinary differential equations with respect to τ subject to initial conditions $x = x_0(s)$, $y = y_0(s)$, $u = u_0(s)$, $p = p_0(s)$, $q = q_0(s)$ at $\tau = 0$, where

$$\frac{du_0}{ds} = p_0\frac{dx_0}{ds} + q_0\frac{dy_0}{ds}$$

and $F(x_0, y_0, u_0, p_0, q_0) = 0$. They have a unique solution for some range of $\tau > 0$, so that x, y, u, p, q can be found in terms of the two parameters s and τ. Moreover, since from eqn (5.44)

$$\frac{\partial F}{\partial \tau} = \frac{\partial F}{\partial x}\frac{\partial x}{\partial \tau} + \frac{\partial F}{\partial y}\frac{\partial y}{\partial \tau} + \frac{\partial F}{\partial u}\frac{\partial u}{\partial \tau} + \frac{\partial F}{\partial p}\frac{\partial p}{\partial \tau} + \frac{\partial F}{\partial q}\frac{\partial q}{\partial \tau} = 0,$$

and at $\tau = 0$, $F = 0$, then x, y, u, p, q satisfy eqn (5.43). In addition

$$\frac{\partial u}{\partial \tau} = p\frac{\partial x}{\partial \tau} + q\frac{\partial y}{\partial \tau}$$

so that $p = \partial u/\partial x$ and $q = \partial u/\partial y$ if

$$\frac{\partial u}{\partial s} = p\frac{\partial x}{\partial s} + q\frac{\partial y}{\partial s}$$

for $\tau > 0$. To verify this define

$$\phi = \frac{\partial u}{\partial s} - p\frac{\partial x}{\partial s} - q\frac{\partial y}{\partial s}$$

so that

$$\frac{\partial \phi}{\partial \tau} = \frac{\partial p}{\partial s}\frac{\partial x}{\partial \tau} + \frac{\partial q}{\partial s}\frac{\partial y}{\partial \tau} - \frac{\partial p}{\partial \tau}\frac{\partial x}{\partial s} - \frac{\partial q}{\partial \tau}\frac{\partial y}{\partial s} = \frac{\partial F}{\partial s} - \phi\frac{\partial F}{\partial u}.$$

Since $F = 0$ for all τ and any s, $\partial F/\partial s = 0$; also $\partial F/\partial u$ is a function of s and τ (and not of ϕ) and $\phi = 0$ at $\tau = 0$. Hence $\phi = 0$ for all τ and we have verified that the integrals of Charpit's equations do solve the partial differential equation on elimination of s and τ. This solution will break down when the Jacobian $\partial(x, y)/\partial(s, \tau)$ vanishes, and it is a necessary condition on the initial data that it does not vanish when $\tau = 0$.

In the case $\partial F/\partial u = \partial F/\partial x = \partial F/\partial y = 0$, eqn (5.44) reduces to $p = p_0(s)$, $q = q_0(s)$ and $dy/dx = (\partial F/\partial p)/(\partial F/\partial q)$ as in eqn (5.40). In the quasi-linear case when $F = ap + bq + c$, where a, b and c are functions of x, y and u only, eqn (5.44) reduces to

$$\frac{\partial x}{\partial \tau} = a, \quad \frac{\partial y}{\partial \tau} = b, \quad \frac{\partial u}{\partial \tau} = c, \tag{5.45}$$

which are the equations of the characteristic curves as in Chester (1971).

For more than two independent variables the general form may be written $F(\mathbf{x}, \mathbf{p}, u) = 0$. Charpit's equations are now

$$\frac{\partial F}{\partial p_i} = \frac{\partial x_i}{\partial \tau}, \quad \frac{\partial F}{\partial x_i} + p_i \frac{\partial F}{\partial u} = -\frac{\partial p_i}{\partial \tau}, \quad p_i \frac{\partial F}{\partial p_i} = \frac{\partial u}{\partial \tau}, \qquad (5.46)$$

and it may be verified that their integrals do give a parametric solution to the differential equation subject to suitable initial data on $\tau = 0$. However they no longer represent characteristics, since in more than two independent variables the characteristics are surfaces. If they are applied to a far-field problem with dispersion relation $\omega = H(\mathbf{x}, \mathbf{p}, t)$, then F takes the special Hamiltonian form

$$\frac{\partial u}{\partial t} + H(\mathbf{x}, \mathbf{p}, t) = 0. \qquad (5.47)$$

Charpit's equations now give $dt/d\tau = 1$ so that $t = \tau$ and

$$\frac{\partial H}{\partial p_i} = \frac{\partial x_i}{\partial t}, \quad \frac{\partial H}{\partial x_i} = -\frac{\partial p_i}{\partial t}, \qquad (5.48)$$

which are called Hamilton's equations. These equations are of considerable importance in Mechanics, and it is possible to represent solutions of eqn (5.48) by a variational problem which provides a good numerical method for their solution.

With H independent of $\mathbf{x}$, $\mathbf{p}$ is constant along the lines of constant s, which are commonly called *rays* in this context, and they are straight lines at any given time t. Thus for the ship-wave problem where the ship is moving in the negative x direction with speed $V(t)$, eqn (5.41) becomes

$$\frac{\partial g}{\partial t} + V(t) \frac{\partial g}{\partial x} - \left[\left(\frac{\partial g}{\partial x} \right)^2 + \left(\frac{\partial g}{\partial y} \right)^2 \right]^{\frac{1}{4}} = 0. \qquad (5.49)$$

On each ray $\partial g/\partial x$ and $\partial g/\partial y$ are constant, but the rays are in the three-dimensional space (x, y, t) with initial data $x = y = 0$ at $t = t_0$ (see Exercise 5.4a).

Problem V. Seismic detection

In seismic exploration charges are exploded at the surface of the earth and pressure waves are propagated. These waves are partially reflected by strong scatterers beneath the earth's surface, and the reflected signals are recorded at the surface. The position of the scatterers is associated with interfaces between geological layers of different properties, and the problem is to determine the position of the scatterers from the data recorded at the surface.

On geological scales a crude model is to regard the earth as composed of a slightly compressible liquid, so that there is a pressure defined at all points which satisfies the wave equation. This is a much simpler model than if we had regarded the earth as an elastic material, as in Chapter 2, Section 2.5, but both are crude approximations to a complicated situation and we therefore investigate the simpler of the two. A desirable improvement on the simplest form of the wave equation is to take account of the stratification of the earth; that is, we expect properties to vary with depth on geological scales, and we model this by making the speed of sound a function of depth which is slowly varying compared to any wavelength generated by the charge. Hence with z measured vertically downwards from the earth's surface, assumed to be flat, and in a two (space) dimensional situation for simplicity, we consider

$$\frac{\partial^2 p}{\partial t^2} = [c(z)]^2 \left[\frac{\partial^2 p}{\partial x^2} + \frac{\partial^2 p}{\partial z^2} \right]. \tag{V1}$$

The variable speed of sound c, and the necessity for at least two space directions, imply that solutions of eqn (V1) obtained by taking transforms in x and t will have an extremely complicated form in terms of inversion integrals. Moreover, solutions are only required on scales large compared to any wavelength generated by the charge, so we should exploit this fact by looking for far-field solutions. Thus with $p = A(x, z, t)e^{ig(x,z,t)}$, where g is in some sense large,

$$\left(\frac{\partial g}{\partial x} \right)^2 + \left(\frac{\partial g}{\partial z} \right)^2 = \frac{1}{[c(z)]^2} \left(\frac{\partial g}{\partial t} \right)^2, \tag{V2}$$

and A satisfies

$$\frac{\partial g}{\partial x} \frac{\partial A}{\partial x} + \frac{\partial g}{\partial z} \frac{\partial A}{\partial z} - \frac{1}{c^2} \frac{\partial g}{\partial t} \frac{\partial A}{\partial t} + \frac{A}{2} \left(\frac{\partial^2 g}{\partial x^2} + \frac{\partial^2 g}{\partial z^2} - \frac{1}{c^2} \frac{\partial^2 g}{\partial t^2} \right) = 0. \tag{V3}$$

Here c varies with z on the far-field length scale, and the eqns (5.46) valid along a ray are

$$\frac{\partial x}{\partial \tau} = g_x, \quad \frac{\partial z}{\partial \tau} = g_z, \quad \frac{\partial t}{\partial \tau} = -\frac{1}{c^2} g_t,$$

$$\frac{\partial g}{\partial \tau} = \frac{\partial g_x}{\partial \tau} = \frac{\partial g_t}{\partial \tau} = 0, \quad \frac{\partial g_z}{\partial \tau} = -\frac{c_z}{c^3} g_t^2,$$

(V4)

where subscripts are used to denote derivatives where convenient and τ is twice that used in eqns (5.46).

The simplification in eqn (V2) for the far field (or high-frequency oscillation) of the wave equation was first used in the problem of light propagation, and is called the *geometric optics approach*. Light intensity, or in our case pressure, is propagated along the rays in a manner analogous to that of particle motion. Because the medium does not have constant properties the rays are not straight lines, and when a ray meets a strong scatterer it is reflected. The crucial simplification of this approach is that we can 'follow the particle' as τ increases, and the existence of a scatterer or non-uniformity at $\tau = \tau_0$ does not influence the solution in $\tau < \tau_0$. The problem can therefore be separated into sections: before and after meeting a scatterer, and near the scatterer itself.

First consider the problem of a single charge exploded at $t = 0$ at $x = 0 = z$ as in Fig. V1 with a single scatterer $z = h(x)$. Data is received at a typical point X where $x = \xi$ after a time $T(\xi)$, and the problem is to determine the scatterer shape $z = h(x)$ from a knowledge of $T(\xi)$. It is therefore an inverse problem, and to appreciate the difficulties we consider the case in which c is

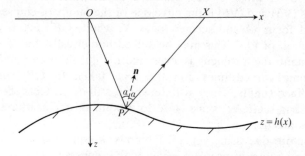

Fig. V1. Ray diagram for scattering from a point source

constant. Equations (V4) then integrate to give

$$x = g_x \tau, \quad z = g_z \tau, \quad t = -\frac{\tau}{c^2} g_t$$

where g_x, g_z, g_t are constant on the rays which are therefore straight lines. Substituting in eqn (V2), $c^2 t^2 = x^2 + z^2$, and the speed of propagation is of course the wave speed c. Hence the time taken for the signal to travel to X is $OP/c + PX/c$, and $OP + PX = cT(\xi)$. Thus for a given ξ, P lies on an ellipse with foci O and X, its actual position being determined by conditions at the scatterer.

Near the scatterer we have an incident periodic wave of speed c making an angle α with the normal $\boldsymbol{n}$ to the scatterer, together with a reflected wave. If we assume that on a strong scatterer the disturbance must vanish, and there is no transmitted wave, then there has to be perfect reflection. Thus the reflected wave also makes an angle α with the normal, leading to the complicated condition

$$\frac{xh' - h}{(x^2 + h^2)^{\frac{1}{2}}} = \frac{(\xi - x)h' + h}{[(\xi - x)^2 + h^2]^{\frac{1}{2}}} \tag{V5}$$

where ξ satisfies

$$(x^2 + h^2)^{\frac{1}{2}} + [(\xi - x)^2 + h^2]^{\frac{1}{2}} = cT(\xi). \tag{V6}$$

If ξ can be eliminated between eqns (V5) and (V6), a highly non-linear first-order differential equation for h will be obtained. However, since chords from the foci of an ellipse are equi-inclined to the normal at all points, the scatterer must touch the ellipse on which P lies. Hence a geometrical interpretation of eqns (V5) and (V6) is the envelope of the family of ellipses with a fixed focus O and second focus X, where $OP + PX$ is a given function of OX. This may be solved numerically, but if c is not constant the problem is even more difficult because the rays become curved lines and the time taken to OP cannot be obtained simply. There will, however, still be perfect reflection at a strong scatterer, since close to a point on the scatterer conditions are not varying.

Before discussing general integrals of eqn (V4) we consider a second common problem, again with c constant. In this problem a series of charges at different points on the surface are fired at

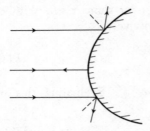

Fig. V2. Ray diagram for scattering from a plane wave

$t = 0$ and signals recorded at later times on the surface. When a plane wave is incident on a scatterer as in Fig. V2, the reflected signal is dispersed into a spectrum of directions and its amplitude will vary with direction, leading to the idea of a *scattering cross-section*. The maximum amplitude of the reflected wave will occur for normal reflection and the scattering cross-section may often be narrow, that is the amplitude of the reflected waves decreases rapidly away from normal reflection (see Exercise 5.4b). Thus the assumption is made that the dominant contribution to the received signal comes only from the normal reflection of the rays, as in Fig. V3. In the case of constant c, the point P of reflection must then be on a circle centre X, where $x = \xi$, and radius $\frac{1}{2}cT(\xi)$, and the scatterer is the envelope of these circles. Note that there is not necessarily a one-to-one relation between X and P.

The problem may be described analytically by considering only the return trajectory PX with initial data $\tau = 0$, $t = \frac{1}{2}T(\xi)$, $x = \eta$, $z = h(\eta)$. This trajectory has to pass through the point $x = \xi$,

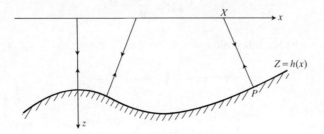

Fig. V3. Ray diagram for sequence of point sources and normal reflection

$z = 0$ when $t = T(\xi)$ for some $\tau > 0$. With c constant eqn (V4) integrates simply to give

$$x - \eta = \tau g_x, \quad z - h(\eta) = \tau g_z, \quad t - \tfrac{1}{2}T(\xi) = -\frac{\tau}{c^2} g_t,$$

where $g_x^2 + g_z^2 = (1/c^2)g_t^2$. Evaluating this on the surface and eliminating τ,

$$(\xi - \eta)^2 + [h(\eta)]^2 = \frac{c^2 T^2}{4}. \tag{V7}$$

Since the ray is normal to the scatterer,

$$\xi - \eta = hh'$$

and eliminating ξ gives a non-linear first-order differential equation for $h(\eta)$.

If the scatterer is a line source (in the y direction) then η is fixed, say equal to zero, and $h = h_0$. Equation (V7) gives the signal response from the sequence of waves due to different charges which are all reflected from the line scatterer. This is called the *migration hyperbola*, and many empirical and numerical techniques are used in the oil industry to approximate a known $T(\xi)$ by a distribution of superposed line scatterers.

This is the simplest solution in which to attempt to introduce a variable speed of sound. With $c = c(z)$ we can integrate eqn (V4) to obtain g_x and g_t constant on the curved rays, and g_z defined by eqn (V2) in terms of z. A further integral is $x = \tau g_x$, and

$$\frac{dz}{d\tau} = -\frac{1}{c}(g_t^2 - c^2 g_x^2)^{\frac{1}{2}}, \quad \frac{dt}{dz} = \frac{g_t}{c}(g_t^2 - c^2 g_x^2)^{-\frac{1}{2}}.$$

Thus further integrals are

$$\tau = \frac{x}{g_x} = -\int_{h_0}^{z} \frac{c\, dz}{(g_t^2 - c^2 g_x^2)^{\frac{1}{2}}},$$

and

$$t - \tfrac{1}{2}T(\xi) = g_t \int_{h_0}^{z} \frac{dz}{c(g_t^2 - c^2 g_x^2)^{\frac{1}{2}}}.$$

If we evaluate these integrals on $z = 0$ where $x = \xi$ and $t = T(\xi)$, and also write $g_t = -kg_x$, then

$$\xi = \int_0^{h_o} \frac{c\,dz}{(k^2-c^2)^{\frac{1}{2}}}, \quad T(\xi) = 2k \int_0^{h_o} \frac{dz}{c(k^2-c^2)^{\frac{1}{2}}}. \tag{V8}$$

The migration hyperbola then generalizes to the eliminant of k between the two relations in eqn (V8).

For the problem of perfect reflection from a general scatterer $z = h(x)$, eqns (V8) are replaced by

$$\xi - \eta = \int_0^{h(\eta)} \frac{c\,dz}{(k^2-c^2)^{\frac{1}{2}}}, \quad T(\xi) = 2k \int_0^{h(\eta)} \frac{d\,z}{c(k^2-c^2)^{\frac{1}{2}}}. \tag{V9}$$

The slope of the ray at the scatterer may be obtained from $dz/dx = g_z/g_x$, so that for it to be normal to the scatterer $h'(\eta)g_z + g_x = 0$ and hence

$$h'(\eta)[k^2 - c^2(h)]^{\frac{1}{2}} = c(h). \tag{V10}$$

The solution is obtained by eliminating ξ and k between the three relations of eqns (V9) and (V10). If $c(z)$ is given such that the integrals in eqn (V9) can be evaluated in closed form, for example as a linear function of z, then k and ξ can always be eliminated to give a non-linear first-order equation for $h(\eta)$. Such solutions, even if not very relevant to real data for $c(z)$ and $T(\xi)$, can give valuable checks on numerical methods.

The problem of a single charge and perfect reflection at each point of the scatterer may also be generated in an obvious way, but both the incoming trajectory OP and outgoing trajectory PX of Fig. V1 have to be considered separately and the times taken added together. This will be very complicated, and a numerical procedure is much to be preferred. One simple case which can be investigated analytically is the direct problem of finding $T(\xi)$ when the scatterer is a horizontal plane, or when $c(z)$ is piecewise constant which is discussed in Exercise 5.4c.

Exercises

5.1. (a) The function $v(t)$ satisfies

$$\varepsilon \frac{dv}{dt} = av(t - \varepsilon t) + bv(t), \quad v(0) = 1,$$

where $a < 0$, $b < 0$. Show that a possible asymptotic expression for small ε is

$$v \sim \exp\left[\frac{1}{\varepsilon} \int_0^t g(s)\, ds\right]$$

provided $0 < t < t_0$, where t_0 is the real root of

$$ate^{1-bt} + 1 = 0,$$

and g satisfies

$$g = ae^{-gt} + b.$$

For $t > t_0$ look for a complex solution $g = (F + iG)/t$ and show that, for large t, $F \sim \log|a/b|$ and $G \sim \pm\pi$. Discuss how this might be used to show that for large t

$$v \sim ct^{(1/\varepsilon)\log|a/b|} \cos\left(\frac{\pi}{\varepsilon} \log t + \alpha\right),$$

where c and α are constants.

Comment on the difference in qualitative behaviour between the cases $a < 0$ and $a > 0$. (Full details may be found in Fox, Mayers, Ockendon and Tayler (1971)).

(b) Consider the problem

$$\varepsilon \frac{d^2u}{dx^2} + x \frac{du}{dx} - xu = 0; \quad u(0) = 0, \quad u(1) = e,$$

and show that there is a boundary layer at $x = 0$ of thickness $O(\varepsilon^{\frac{1}{2}})$. If a solution is required correct to $O(\varepsilon)$, show that the inner solution in the boundary layer must contain terms of $O(\varepsilon \log \varepsilon)$ in addition to powers of $\varepsilon^{\frac{1}{2}}$. Discuss the matching in detail and show that this term is

$$-\frac{\varepsilon \log \varepsilon}{(2\pi)^{\frac{1}{2}}} \int_0^z e^{-t^2/2}\, dt$$

where $x = \varepsilon^{\frac{1}{2}} z$.

(c) Show that a uniformly valid solution for the Van der Pol oscillator modelled by

$$\frac{d^2x}{dt^2} + x = \varepsilon \frac{dx}{dt}(\alpha - \beta x^2),$$

can be found in the form

$$x = a(\tau)\sin t + b(\tau)\cos t, \quad \tau = \varepsilon t,$$

where a and b satisfy first-order ordinary differential equations. If at $t = 0$, $x = 0$, $\dot{x} = 1$, show that this solution has the form

$$x = \left[\frac{4\alpha}{(4\alpha - \beta)e^{-\varepsilon \alpha t} + \beta} \right]^{\frac{1}{2}} \sin t.$$

Describe the various possibilities after a large time for positive and negative values of α and β.

5.2. (a) A two-dimensional bearing consists of a flat plate pivoted at its mid-point and a plane boundary moving with constant speed in its plane. Show that in suitable non-dimensional coordinates

$$p = \frac{1}{n} \left(\frac{6}{1 - nx} - \frac{3(1 - n^2)}{(1 - nx)^2} - 3 \right)$$

where the slope of the plate relative to the plane boundary is $-n$ ($0 < n \ll 1$). Show also that the anticlockwise moment on the plate is given by $\int_{-1}^{1} xp \, dx +$ smaller terms, and that the centre of pressure is at $x = 6n/5$.

If the bearing is freely pivoted at a general point, discuss whether a stable equilibrium position is possible.

(b) A heated circular cylinder of radius a is pushed with speed U through wax which melts in a thin layer in front of the cylinder. Show that the thickness of the layer is $O(a \, Ste/Pe)$ where the Stefan number $Ste = cT_0/L \ll 1$, and T_0 is the cylinder temperature above the wax melting temperature (c is the molten wax specific heat and L the latent heat of melting). If θ is the angle measured at the centre of the cylinder such that $\theta = 0$ is the direction of motion, show that a lubrication analysis for the molten layer leads to

$$u = \frac{6 \sin \theta}{\delta^3} z(z - \delta)$$

where u is the transverse velocity and z is measured radially across the layer whose thickness is $\delta(\theta)$.

Show that if $\lambda = U^4 L^2 \rho^2 \mu a^2 / k^3 T_0^3$ is $O(1)$ then heat dissipation effects dominate convective effects in the layer, so that the non-dimensional layer temperature (above melting) is given by

$$T = 1 - \frac{z}{\delta} - \frac{6\lambda \sin^2\theta}{\delta^6}(2z^4 - 4z^3\delta + 3z^2\delta^2 - z\delta^3).$$

Given that the Stefan condition reduces to $\partial T/\partial z = -\cos\theta$, $z = \delta$ show that the layer thickness is determined from

$$\delta^3\cos\theta - \delta^2 - 6\lambda \sin^2\theta = 0.$$

(c) In Problem T an alternative model for the paper is a porous material for which which Darcy's law holds and whose porosity depends on the compression of the paper. Treating the paper as a thin layer, derive an alternative model to eqn (T3) in the form

$$\frac{d}{dx}\left(h - \frac{h^3}{12}\frac{dp}{dx}\right) = (s_0 - s)\frac{d}{dx}\left[\phi(s_0 - s)\frac{dp}{dx}\right].$$

For small compressibility of the paper show that this leads to a non-linear first-order equation for p, valid for all x.

5.3. (a) Show that the transformation from (x, y) to (x, ψ) variables reduces the boundary-layer equation (5.19) for flow past a flat plate to

$$\frac{\partial u}{\partial x} = \frac{\partial}{\partial \psi}\left(u\frac{\partial u}{\partial \psi}\right), \quad \text{where} \quad u = \frac{\partial \psi}{\partial y}.$$

Show that a similarity solution $u = F(\eta)$, where $\eta = \psi/x^{\frac{1}{2}}$, exists where F satisfies

$$FF'' + (F')^2 + \tfrac{1}{2}\eta F' = 0, \quad F(0) = 0, \quad F(\infty) = 1.$$

Describe an appropriate solution in the phase plane, following the discussion of eqn (4.40).

(b) Given that the inviscid flow past a wedge has a pressure distribution $cx^2/2$ on the wedge, where x is measured along the wedge surface, show that the boundary-layer equations (5.18) may be reduced by a similarity transformation to the problem

$$f''' + f'' + 1 - (f')^2 = 0, \quad f(0) = 0 = f'(0), \quad f'(\infty) = 1.$$

(c) A one-dimensional mould is filled by a constant inlet

pressure p_0, and the moving front $x = s(t)$ reaches the end wall $x = 1$ at time t_0. If the molten material is slightly compressible, and assuming a linear relation between small density and pressure changes, obtain an estimate for the time scale of the packing stage. Show that in this stage the continuity equation (U1) has to be modified, and that with a constant viscosity the pressure satisfies

$$\frac{\partial^2 p}{\partial x^2} = 12 \frac{\partial p}{\partial \tau}; \quad x = 0, \quad p = p_0; \quad x = 1, \quad \frac{\partial p}{\partial x} = 0;$$

$$\tau = 0, \quad p = p_0(1 - x).$$

Obtain a solution for p in terms of a Fourier expansion and discuss the limitations of the model.

5.4. (a) A point source is accelerating in the negative x direction with constant acceleration c on the surface of a sea of infinite depth. Show that in a frame moving with the source the far-field solution $\phi = A \mathrm{e}^{\mathrm{i} g(x,y,t)}$ may be determined parametrically from

$$x = \frac{c}{2}(t^2 - t_0^2) - (t - t_0)\frac{\cos\theta}{2k^{\frac{1}{2}}},$$

$$y = -(t - t_0)\frac{\sin\theta}{2k^{\frac{1}{2}}}, \quad ct_0 k^{\frac{1}{2}} \cos\theta = 1,$$

and

$$g = -\frac{c}{2}(t^2 - t_0^2)k \cos\theta + k^{\frac{1}{2}}(t - t_0).$$

Eliminate k, and with $\alpha = t_0/t$ define $p(\alpha, \theta)$ and $q(\alpha, \theta)$ such that

$$\frac{x}{t^2} = p(\alpha, \theta), \quad \frac{y}{t^2} = q(\alpha, \theta).$$

Hence determine the envelope of the rays in the form

$$\frac{x}{t^2} = \left(\frac{c}{2}\right)\frac{(1 - \alpha)(1 + 3\alpha)}{(1 + 2\alpha)}, \quad \frac{y}{t^2} = \mp\left(\frac{c}{\sqrt{2}}\right)\frac{(1 - \alpha)\alpha^{\frac{3}{2}}}{(1 + 2\alpha)}, \quad 0 < \alpha < 1.$$

Describe the disturbance region created by the point source accelerating from the rest at time $t = 0$.

(b) With a two-dimensional wave equation and constant wave-speed equal to unity, a plane wave of unit amplitude and frequency ω is incident from the positive x direction on a scatterer defined by $x = x_0(s)$, $y = y_0(s)$. Show that in the high-frequency approximation the reflected wave is $\psi = A(x, y)e^{i\omega g(x,y)}$, where $(\partial g/\partial x)^2 + (\partial g/\partial y)^2 = 1$ and $g = x_0(s)$, $A = 1$ at the scatterer. Hence show that the slope of each reflected ray is $\tan \theta$ where $y_0'/x_0' = \tan(\theta/2)$, and deduce that reflected and incident rays are equi-inclined to the normal to the scatterer.

If t is a variable along each ray show that, using eqn (V3),

$$\frac{\partial A}{\partial t} + A\left(p_0'\frac{\partial s}{\partial x} + q_0'\frac{\partial s}{\partial y}\right) = 0.$$

By differentiating x and y determine $\partial s/\partial x$ and $\partial s/\partial y$ in terms of p_0, q_0, x_0, y_0 and their derivatives, and hence obtain

$$\frac{1}{A^2} = \frac{2t(p_0'q_0 - p_0q_0') + q_0x_0' - p_0y_0'}{q_0x_0' - p_0y_0'}.$$

Show that as $|x^2 + y^2| \to \infty$, $A^2 \simeq F(s)/|x^2 + y^2|$ where

$$F(s) = \frac{q_0x_0' - p_0y_0'}{p_0'q_0 - q_0'p_0}$$

is the scattering cross-section.

(c) Show that at an interface $z = f(x)$ between two media, with speeds of sound c_1 and c_2, each ray is refracted such that $g_x + f'g_z$ and $c(g_x^2 + g_z^2)^{\frac{1}{2}}$ are continuous. Verify that this is equivalent to Snell's law $\sin \theta_1/c_1 = \sin \theta_2/c_2$, where θ_1 and θ_2 are the angles the ray makes with the normal to the interface.

For a horizontal layering, so that the speed of sound is c_2 in $0 < z < f_0$ and c_1 in $z > f_0$, show that the line-source scatterer at $x = 0$, $z = h_0 > f_0$ gives a 'migration curve' defined by

$$T(\xi) = \frac{[x^2 + (h_0 - f_0)^2]^{\frac{1}{2}}}{c_1} + \frac{[(\xi - x)^2 + f_0^2]^{\frac{1}{2}}}{c_2},$$

where x is to be eliminated from the relation

$$c_1^2(\xi - x)^2[x^2 + (h_0 - f_0)^2] = c_2^2x^2[(\xi - x)^2 + f_0^2].$$

Bibliography

AITCHISON, J. M., ELLIOTT, C. M., and OCKENDON, J. R. (1983). Percolation in gently sloping beaches. *IMA J. Appl. Maths* **30**, 269–288.

AITCHISON, J. M., LACEY, A. A., and SHILLOR, M. (1984). A model for an electropaint process. *IMA J. Appl. Maths* **33**, 17–32.

ATTHEY, D. R. (1974). A finite difference scheme for melting problems. *J. Inst. Math. Appl.* **13**, 353–366.

BENDER, C. M. and ORSZAG, S. A. (1978). *Advanced mathematical methods for scientists and engineers.* McGraw-Hill, New York.

BIRKHOFF, G. and ZARANTONELLO, E. H. (1957). *Jets, wakes and cavities.* Academic Press, New York.

BRENNAN, C. and PEARCE, J. C. (1978). Granular material flow in two-dimensional hoppers. *J. Appl. Mech.* **45**, 43–50.

CAMERON, A. (1981). *Basic lubrication theory.* Ellis–Horwood, Chichester.

CHALMERS, B. (1964). *Principles of solidification.* Wiley, New York.

CHESTER, C. R. (1971). *Techniques in partial differential equations.* McGraw-Hill, New York.

COURANT, R. and HILBERT, D. (1976). *Methods of mathematical physics,* vol. I. Interscience, New York.

CRANK, J. (1984) *Free and moving boundary problems.* Clarendon Press, Oxford.

DOWDEN, J., DAVIS, M., and KAPADIA, P. (1983). Some aspects of the fluid dynamics of laser welding. *J. Fluid Mech.* **126**, 123–146.

ELLIOTT, C. M. and OCKENDON, J. R. (1982). *Weak and variational methods for moving boundary problems.* Pitman, London.

ENGLAND, A. M. (1971). *Complex variable methods in elasticity.* Interscience, London.

FITT, V., OCKENDON, J. R., and SHILLOR, M. (1985). Counter current mass transfer. *Int. J. Heat Mass Trans.* In press.

FOX, L., MAYERS, D. F., OCKENDON, J. R., and TAYLER, A. B. (1971). On a functional differential equation. *J. Inst. Math. Appl.* **8**, 271–307.

GAKHOV, F. D. (1966). *Boundary value problems.* Pergamon, Oxford.

GRAFF, K. F. (1975). *Wave motion in elastic solids.* Clarendon Press, Oxford.

HILDEBRAND, F. B. (1975). *Methods of applied mathematics.* McGraw-Hill, New York.

JEFFREY, A. and KAWAHARA, T. (1982). *Asymptotic methods in non-linear wave theory.* Pitman, London.

JORDAN, D. W. and SMITH, P. (1977). *Nonlinear ordinary differential equations.* Clarendon Press, Oxford.

LACEY, A. A., OCKENDON, J. R., and TAYLER, A. B. (1982). "Waiting time" solutions of a nonlinear diffusion equation. *SIAM J. Appl. Math.* **42**, 1252–1264.

LACEY, A. A. and SHILLOR, M. (1983). The existence and stability of regions with superheating in the classical two-phase one-dimensional Stefan problem with heat sources. *IMA J. Appl. Math.* **30**, 215–230.

LACEY, A. A. and TAYLER, A. B. (1983). A mushy region in a Stefan problem. *IMA J. Appl. Math.* **30**, 303–313.

LANDAU, L. D. and LIFSHITZ, E. M. (1963). *Fluid mechanics.* Pergamon, London.

LAX, P. D. (1973). *Hyperbolic systems of conservation laws and the mathematical theory of shock waves.* SIAM, Philadelphia.

LIN, C. C. and SEGAL, L. A. (1974). *Mathematics applied to deterministic problems in the natural sciences.* Macmillan, New York.

LOVE, A. E. H. (1944). *A treatise on the mathematical theory of elasticity.* Dover, New York.

MACKIE, A. G. (1965). *Boundary value problems.* Oliver and Boyd, Edinburgh.

MACLAINE-CROSS, I. L. and BANKS, P. J. (1972). Coupled heat and mass transfer regenerators—prediction using an analogy with heat transfer. *Int. J. Heat Mass Trans.* **15**, 1225–1241.

MURRAY, J. D. (1984). *Asymptotic analysis.* Springer-Verlag, New York.

NEUMAN, S. P. (1977). Theoretical derivation of Darcy's Law. *Acta Mech.* **25**, 153–170.

NOBLE, B. (1958). *Methods based on the Wiener-Hopf technique for the solution of partial differential equations.* Pergamon, London.

OCKENDON, J. R. and TAYLER, A. B. (1971). The dynamics of a current collection system for an electric locomotive. *Proc. Roy. Soc.* A **322**, 447–468.

OCKENDON, H. (1979). Channel flow with temperature dependent viscosity and internal viscous dissipation. *J. Fluid Mech.* **93**, 737–746.

OCKENDON, H. and TAYLER, A. B. (1983). *Inviscid fluid flows.* Springer-Verlag, New York.

PIPKIN, A. C. (1972). *Lectures on visco-elasticity theory.* Springer-Verlag, New York.

POLUBARINOVA-KOCHINA, P. Y. (1962). *Theory of ground water movement.* University Press, Princeton.

PROTTER, M. H. and WEINBERGER, H. F. (1967). *Maximum principles in differential equations.* Prentice-Hall, Englewood Cliffs.

RODEMAN, R., LONGCOPE, D. B., and SHAMPINE, L. F. (1976). Response of a string to an accelerating mass. *SIAM J. Appl. Mech.* **98,** 675–679.

ROMIE, F. E. (1979). Periodic thermal storage: the regenerator. *Trans. ASME J. Heat Trans.* **101,** 727–731.

RUBINSTEIN, L. I. (1971). *The Stefan problem.* Amer. Math. Soc., Providence.

SAFFMAN, P. G. and TAYLOR, G. I. (1958). The penetration of fluid into a porous medium or Hele–Shaw cell. *Proc. Roy. Soc.* A **245,** 312–329.

SNEDDON, I. N. (1951). *Fourier transforms,* McGraw-Hill, New York.

STAKGOLD, I. (1979). *Green's functions and boundary value problems.* Wiley-Interscience, New York.

TAYLER, A. B. and NICHOLAS, M. O. (1982). Unsteady flows over a cooled flat plate. *IMA J. Appl. Math.* **28,** 75–91.

VAN DYKE, M. (1964). *Perturbation methods in fluid mechanics.* Academic Press, New York.

WOODRUFF, D. P. (1973). *The solid–liquid interface.* University Press, Cambridge.

Index